PROBLEMS FOR STUDENT INVESTIGATION

Michael B. Jackson and John R. Ramsay, Editors

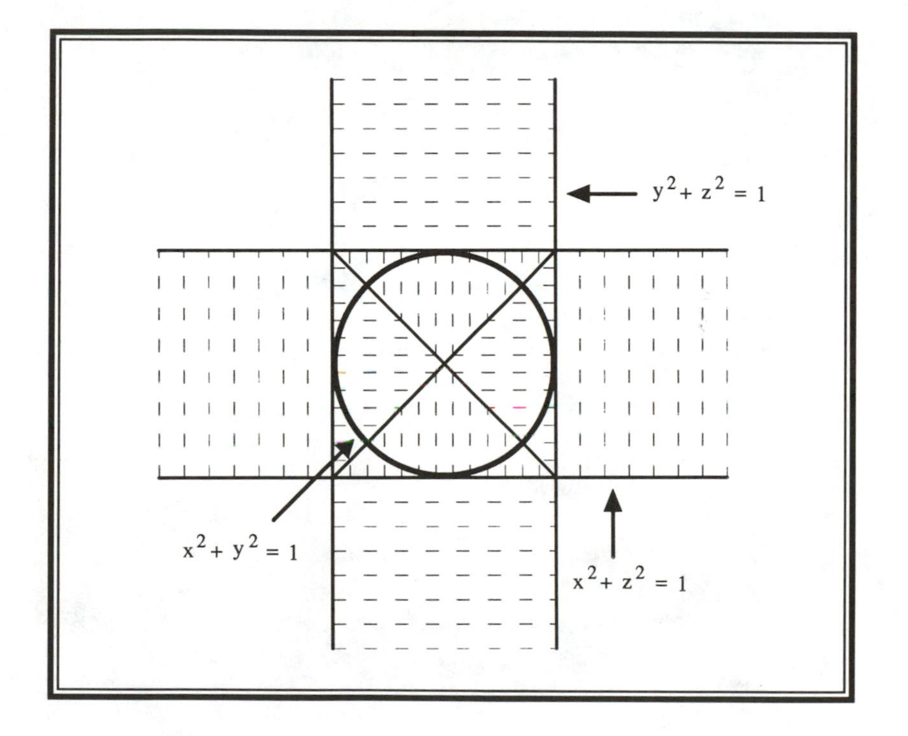

Resources for Calculus Collection

Volume 4

PROBLEMS FOR STUDENT INVESTIGATION

Michael B. Jackson and John R. Ramsay, Editors

A Project of
The Associated Colleges of the Midwest and
The Great Lakes Colleges Association

Supported by the National Science Foundation
A. Wayne Roberts, Project Director

MAA Notes Volume 30

Published and Distributed by
The Mathematical Association of America

MAA Notes and Reports Series

The MAA Notes and Reports Series, started in 1982, addresses a broad range of topics and themes of interest to all who are involved with undergraduate mathematics. The volumes in this series are readable, informative, and useful, and help the mathematical community keep up with developments of importance to mathematics.

MAA Notes

1. Problem Solving in the Mathematics Curriculum, *Committee on the Teaching of Undergraduate Mathematics*, a subcommittee of the Committee on the Undergraduate Program in Mathematics, *Alan H. Schoenfeld*, Editor
2. Recommendations on the Mathematical Preparation of Teachers, *Committee on the Undergraduate Program in Mathematics, Panel on Teacher Training*.
3. Undergraduate Mathematics Education in the People's Republic of China, *Lynn A. Steen*, Editor.
5. American Perspectives on the Fifth International Congress on Mathematical Education, *Warren Page*, Editor.
6. Toward a Lean and Lively Calculus, *Ronald G. Douglas*, Editor.
8. Calculus for a New Century, *Lynn A. Steen*, Editor.
9. Computers and Mathematics: The Use of Computers in Undergraduate Instruction, *Committee on Computers in Mathematics Education, D. A. Smith, G. J. Porter, L. C. Leinbach, and R. H. Wenger*, Editors.
10. Guidelines for the Continuing Mathematical Education of Teachers, *Committee on the Mathematical Education of Teachers*.
11. Keys to Improved Instruction by Teaching Assistants and Part-Time Instructors, *Committee on Teaching Assistants and Part-Time Instructors, Bettye Anne Case*, Editor.
13. Reshaping College Mathematics, *Committee on the Undergraduate Program in Mathematics, Lynn A. Steen*, Editor.
14. Mathematical Writing, by *Donald E. Knuth, Tracy Larrabee, and Paul M. Roberts*.
15. Discrete Mathematics in the First Two Years, *Anthony Ralston*, Editor.
16. Using Writing to Teach Mathematics, *Andrew Sterrett*, Editor.
17. Priming the Calculus Pump: Innovations and Resources, *Committee on Calculus Reform and the First Two Years*, a subcomittee of the Committee on the Undergraduate Program in Mathematics, *Thomas W. Tucker*, Editor.
18. Models for Undergraduate Research in Mathematics, *Lester Senechal*, Editor.
19. Visualization in Teaching and Learning Mathematics, *Committee on Computers in Mathematics Education, Steve Cunningham and Walter S. Zimmermann*, Editors.
20. The Laboratory Approach to Teaching Calculus, *L. Carl Leinbach et al.*, Editors.
21. Perspectives on Contemporary Statistics, *David C. Hoaglin and David S. Moore*, Editors.
22. Heeding the Call for Change: Suggestions for Curricular Action, *Lynn A. Steen*, Editor.
23. Statistical Abstract of Undergraduate Programs in the Mathematical Sciences and Computer Science in the United States: 1990–91 CBMS Survey, *Donald J. Albers, Don O. Loftsgaarden, Donald C. Rung, and Ann E. Watkins*.
24. Symbolic Computation in Undergraduate Mathematics Education, *Zaven A. Karian*, Editor.
25. The Concept of Function: Aspects of Epistemology and Pedagogy, *Guershon Harel and Ed Dubinsky*, Editors.

26. Statistics for the Twenty-First Century, *Florence and Sheldon Gordon,* Editors.
27. Resources for Calculus Collection, Volume 1: Learning by Discovery: A Lab Manual for Calculus, *Anita E. Solow,* Editor.
28. Resources for Calculus Collection, Volume 2: Calculus Problems for a New Century, *Robert Fraga,* Editor.
29. Resources for Calculus Collection, Volume 3: Applications of Calculus, *Philip Straffin,* Editor.
30. Resources for Calculus Collection, Volume 4: Problems for Student Investigation, *Michael B. Jackson and John R. Ramsay,* Editors.
31. Resources for Calculus Collection, Volume 5: Readings for Calculus, *Underwood Dudley,* Editor.

MAA Reports

1. A Curriculum in Flux: Mathematics at Two-Year Colleges, *Subcommittee on Mathematics Curriculum at Two-Year Colleges,* a joint committee of the MAA and the American Mathematical Association of Two-Year Colleges, *Ronald M. Davis,* Editor.
2. A Source Book for College Mathematics Teaching, *Committee on the Teaching of Undergraduate Mathematics, Alan H. Schoenfeld,* Editor.
3. A Call for Change: Recommendations for the Mathematical Preparation of Teachers of Mathematics, *Committee on the Mathematical Education of Teachers, James R. C. Leitzel,* Editor.
4. Library Recommendations for Undergraduate Mathematics, *CUPM ad hoc Subcommittee, Lynn A. Steen,* Editor.
5. Two-Year College Mathematics Library Recommendations, *CUPM ad hoc Subcommittee, Lynn A. Steen,* Editor.

These volumes may be ordered from the Mathematical Association of America, 1529 Eighteenth Street, NW, Washington, DC 20036.
202-387-5200 FAX 202-265-2384

First Printing

ISBN 0-88385-086-9
Library of Congress Catalog Number 92-62282
Printed in the United States of America
Current Printing
10 9 8 7 6 5 4 3 2 1

INTRODUCTION
RESOURCES FOR CALCULUS COLLECTION

Beginning with a conference at Tulane University in January, 1986, there developed in the mathematics community a sense that calculus was not being taught in a way befitting a subject that was at once the culmination of the secondary mathematics curriculum and the gateway to collegiate science and mathematics. Far too many of the students who started the course were failing to complete it with a grade of C or better, and perhaps worse, an embarrassing number who did complete it professed either not to understand it or not to like it, or both. For most students it was not a satisfying culmination of their secondary preparation, and it was not a gateway to future work. It was an exit.

Much of the difficulty had to do with the delivery system: classes that were too large, senior faculty who had largely deserted the course, and teaching assistants whose time and interest were focused on their own graduate work. Other difficulties came from well intentioned efforts to pack into the course all the topics demanded by the increasing number of disciplines requiring calculus of their students. It was acknowledged, however, that if the course had indeed become a blur for students, it just might be because those choosing the topics to be presented and the methods for presenting them had not kept their goals in focus.

It was to these latter concerns that we responded in designing our project. We agreed that there ought to be an opportunity for students to discover instead of always being told. We agreed that the availability of calculators and computers not only called for exercises that would not be rendered trivial by such technology, but would in fact direct attention more to ideas than to techniques. It seemed to us that there should be explanations of applications of calculus that were self-contained, and both accessible and relevant to students. We were persuaded that calculus students should, like students in any other college course, have some assignments that called for library work, some pondering, some imagination, and above all, a clearly reasoned and written conclusion. Finally, we came to believe that there should be available to students some collateral readings that would set calculus in an intellectual context.

We reasoned that the achievement of these goals called for the availability of new materials, and that the uncertainty of just what might work, coupled with the number of people trying to address the difficulties, called for a large collection of materials from which individuals could select. Our goal was to develop such materials, and to encourage people to use them in any way they saw fit. In this spirit, and with the help of the Notes editor and committee of the Mathematical Association of America, we have produced five volumes of materials that are, with the exception of volume V where we do not hold original copyrights, meant to be in the public domain.

We expect that some of these materials may be copied directly and handed to an entire class, while others may be given to a single student or group of students. Some will provide a basis from which local adaptations can be developed. We will be pleased if authors ask for permission, which we expect to be generous in granting, to incorporate our materials into texts or laboratory manuals. We hope that in all of these ways, indeed in any way short of reproducing substantial segments to sell for profit, our material will be used to greatly expand ideas about how the calculus might be taught.

Though I as Project Director never entertained the idea that we could write a single text that would be acceptable to all 26 schools in the project, it was clear that some common notion of topics essential to any calculus course would be necessary to give us direction. The task of forging a common syllabus was managed by Andy Sterrett with a tact and efficiency that was instructive to us all, and the product of this work, an annotated core syllabus, appears as an appendix in Volume 1. Some of the other volumes refer to this syllabus to indicate where, in a course, certain materials might be used.

This project was situated in two consortia of liberal arts colleges, not because we intended to develop materials for this specific audience, but because our schools provide a large reservoir of classroom teachers who lavish on calculus the same attention a graduate faculty might give to its introductory analysis course. Our schools, in their totality, were equipped with most varieties of computer labs, and we included in our consortia many people who had become national leaders in the use of computer algebra systems.

We also felt that our campuses gave us the capability to test materials in the classroom. The size of our schools enables us to implement a new idea without cutting through the red tape of a larger institution, and we can just as quickly reverse ourselves when it is apparent that what we are doing is not working. We are practiced in going in both directions. Continual testing of the materials we were developing was seen as an integral part of our project, an activity that George Andrews, with the title of Project Evaluator, kept before us throughout the project.

The value of our contributions will now be judged by the larger mathematical community, but I was right in thinking that I could find in our consortia the great abundance of talent necessary for an undertaking of this magnitude. Anita Solow brought to the project a background of editorial work and quickly became not only one of the editors of our publications, but also a person to whom I turned for advice regarding the project as a whole. Phil Straffin, drawing on his association with UMAP, was an ideal person to edit a collection of applications, and was another person who brought editorial experience to our project. Woody Dudley came to the project as a writer well known for his witty and incisive commentary on mathematical literature, and was an ideal choice to assemble a collection of readings.

Our two editors least experienced in mathematical exposition, Bob Fraga and Mic Jackson, both justified the confidence we placed in them. They brought to the project an enthusiasm and freshness from which we all benefited, and they were able at all points in the project to draw upon an excellent corps of gifted and experienced writers. When, in the last months of the project, Mic Jackson took an overseas assignment on an Earlham program, it was possible to move John Ramsay into Mic's position precisely because of the excellent working relationship that had existed on these writing teams.

The entire team of five editors, project evaluator and syllabus coordinator worked together as a harmonious team over the five year duration of this project. Each member, in turn, developed a group of writers, readers, and classroom users as necessary to complete the task. I believe my chief contribution was to identify and bring these talented people together, and to see that they were supported both financially and by the human resources available in the schools that make up two remarkable consortia.

A. Wayne Roberts
Macalester College
1993

THE FIVE VOLUMES OF THE RESOURCES FOR CALCULUS COLLECTION

1. Learning by Discovery: A Lab Manual for Calculus
Anita E. Solow, editor

The availability of electronic aids for calculating makes it possible for students, led by good questions and suggested experiments, to discover for themselves numerous ideas once accessible only on the basis of theoretical considerations. This collection provides questions and suggestions on 26 different topics. Developed to be independent of any particular hardware or software, these materials can be the basis of formal computer labs or homework assignments. Although designed to be done with the help of a computer algebra system, most of the labs can be successfully done with a graphing calculator.

2. Calculus Problems for a New Century
Robert Fraga, editor

Students still need drill problems to help them master ideas and to give them a sense of progress in their studies. A calculator can be used in many cases, however, to render trivial a list of traditional exercises. This collection, organized by topics commonly grouped in sections of a traditional text, seeks to provide exercises that will accomplish the purposes mentioned above, even for the student making intelligent use of technology.

3. Applications of Calculus
Philip Straffin, editor

Everyone agrees that there should be available some self-contained examples of applications of the calculus that are tractable, relevant, and interesting to students. Here they are, 18 in number, in a form to be consulted by a teacher wanting to enrich a course, to be handed out to a class if it is deemed appropriate to take a day or two of class time for a good application, or to be handed to an individual student with interests not being covered in class.

4. Problems for Student Investigation
Michael B. Jackson and John R. Ramsay, editors

Calculus students should be expected to work on problems that require imagination, outside reading and consultation, cooperation, and coherent writing. They should work on open-ended problems that admit several different approaches and call upon students to defend both their methodology and their conclusion. Here is a source of 30 such projects.

5. Readings for Calculus
Underwood Dudley, editor

Faculty members in most disciplines provide students in beginning courses with some history of their subject, some sense not only of what was done by whom, but also of how the discipline has contributed to intellectual history. These essays, appropriate for duplicating and handing out as collateral reading aim to provide such background, and also to develop an understanding of how mathematicians view their discipline.

ACKNOWLEDGEMENTS

Besides serving as editors of the collections with which their names are associated, Underwood Dudley, Bob Fraga, Mic Jackson, John Ramsay, Anita Solow, and Phil Straffin joined George Andrews (Project Evaluator), Andy Sterrett (Syllabus Coordinator) and Wayne Roberts (Project Director) to form a steering committee. The activities of this group, together with the writers' groups assembled by the editors, were supported by two grants from the National Science Foundation.

The NSF grants also funded two conferences at Lake Forest College that were essential to getting wide participation in the consortia colleges, and enabled member colleges to integrate our materials into their courses.

The projects benefited greatly from the counsel of an Advisory Committee that consisted of Morton Brown, Creighton Buck, Jean Callaway, John Rigden, Truman Schwartz, George Sell, and Lynn Steen.

Macalester College served as the grant institution and fiscal agent for this project on behalf of the schools of the Associated Colleges of the Midwest (ACM) and Great Lakes Colleges Association (GLCA) listed below.

ACM

Beloit College
Carleton College
Coe College
Colorado College
Cornell College
Grinnell College
Knox College
Lake Forest College
Lawrence University
Macalester College
Monmouth College
Ripon College
St. Olaf College
University of Chicago

GLCA

Albion College
Antioch College
Denison University
DePauw University
Earlham College
Hope College
Kalamazoo College
Kenyon College
Oberlin College
Ohio Wesleyan University
Wabash College
College of Wooster

I would also like to thank Stan Wagon of Macalester College for providing the cover image for each volume in the collection.

TABLE OF CONTENTS

Preface

In the interest of making calculus more lively, this volume of projects can be used by an instructor to give her students an opportunity to work with a mathematical problem that can be posed easily, but which is impossible for most students to solve as part of an overnight homework assignment. The projects are not intended to be for honors students, but are problems that a small group of typical calculus students can solve given a reasonable amount of time and effort, with some timely guidance from the instructor. Experience indicates that through applying themselves to projects of this kind students develop a better notion of ways in which calculus can be used to solve realistic problems, have the opportunity to look more closely at some of the important concepts of calculus, and gain a sense of personal ownership of some piece of calculus. Some learn how to use the library effectively to find mathematical sources, and all improve their ability to read and write mathematical material and to cooperate with peers in the solution of a difficult problem. Finally, the experience of developing solutions to problems which on the first reading seem inscrutable, increases the confidence of students.

The contents of the projects are distributed over the first year of a typical single-variable calculus program, with some projects applicable to multivariable calculus. Some projects involve the application or extension of a mathematical concept that is part of the content of the usual course, while others give students the opportunity to examine an interesting application or theory somewhat tangential to the core material. Each project is self-contained, including a brief statement of the problem for the students and more thorough information for the instructor. The first item in the information for the instructor is an abstract of the project in which we explain what makes this problem interesting, what we expect students to learn from doing the problem, and some rationale for why we pose the problem as we do. A description of prerequisite knowledge and skills for each project will help instructors determine where that project could fit into a particular course. To further assist in determining proper placement of the projects in your course, a recommended first-semester calculus syllabus is given in outline form with projects of the first three sections of this volume included according to their prerequisite. (An annotated syllabus for Calculus I and II worked out by colleges in the project can be found in the appendix of Volume I of the *Resources in Calculus* set.) Each project description also identifies essential or recommended library or computing resources. The bulk of information for the instructor is a section containing one or more sample solutions. The solutions presented have been chosen to represent approaches students may take on the particular project. They are not necessarily the most concise, most elegant, or even the most intuitive for the instructor.

We are grateful to the many individuals who assisted in this effort. We especially thank the Resources for Calculus steering team for their support and encouragement, and for being such a pleasant group with which to work. Our special thanks to Steve Boyce and Matt Richey for their input in the early stages of the project and to Charlie Jones for his help at "crunch time." Finally, we want to express our deep appreciation to Jackie Middleton for using her technical talents in designing and formatting this volume.

Mic Jackson
Earlham College
1993

John Ramsay
The College of Wooster
1993

Suggestions for Using This Volume

The nature of the projects in this volume is such that one student assigned to do a project alone would often tend to just "get through" and not think deeply nor produce a report of high quality. A small group of students tend to produce better work and show more individual learning. For this reason, we encourage you to assign two or three students to produce a joint report over a period of one to three weeks. Almost normal classroom and homework responsibilities can be continued during the duration of the project, but we have found that the use of at least one class session to work individually with student groups is often appropriate early in the period allotted to the projects. We also recommend that intermediate, extra-class sessions be scheduled during which the instructor can meet briefly with a representative from each group so that person can report the group's progress, ask questions and receive appropriate guidance. For each problem, we have tried to warn the instructor as to where the students tend to have difficulties, and what sort of hints or information might be appropriate to help them move along. We have also included at least one example of a correct way to approach each problem. Some of the projects will naturally lead more imaginative students to attempt general conjectures. For the projects where we would expect this to happen, we have given some advice as to how much an instructor might reasonably expect from an average student, and suggested questions which might challenge more ambitious students.

We expect that most instructors will ask each student group to submit a written report in which they (1) describe the problem as posed, (2) thoroughly explain their solution including all assumptions, interpretations and important calculations, (3) cite any outside resources used, and (4) discuss any related ideas or questions that have been raised by their examination of the problem. If you are not accustomed to grading library research papers, it might be useful to seek assistance from a colleague in the humanities to act as a second reader. You should grade the paper based on its mathematical thoroughness and correctness as well as the quality of writing. If you prefer, you may ask each group to give an oral presentation of their work to their peers instead of, or in addition to, submitting a written report. In either case, it is advisable to have each group submit a draft or outline, including bibliography, at least three or four days before the report is due. You can use these to insure that each group is on track for successful completion.

It is useful to assign individual grades for group projects based on both the quality of the final report and the quality of participation by each individual. One way to determine just how each individual contributed to the final project is to tell the students that each of them will be required to submit a confidential description of her or his contribution (and, possibly, the contributions of each other team member). Another way of encouraging full participation by each team member is to conduct interviews with a randomly chosen individual from each team in which that person must explain the team's work.

Finally, a word of caution. The experience of those who have taken the lead in projects of this nature uniformly report that the expectations of students is a major impediment to change. They have been conditioned, after all, by twelve years experience to think that instruction in mathematics consists of some explanation of "how to work these kinds" followed by a long list of exercises "of these kinds." At least initially, many students resist being required to engage themselves in projects such as these and to work cooperatively. We recommend, therefore, that you anticipate this initial resistance and consider distributing a copy of or some variation of the following "Note to Students."

Note to Students

For many of you a group project will be a new experience. To help you take full advantage of this activity, the following suggestions are offered.

- Get started immediately. You will not be able to complete the project on a last minute basis. Portions of the project will move slowly and working in a group requires more time due to scheduling difficulties.
- Read over the entire project carefully before you begin discussing or completing any portion of it.
- Initially, you may not know how to begin. Don't panic, a discussion with other group members will usually generate some ideas.
- The procedure for solving a project is not as clear cut as it is for solving standard homework problems. You will possibly need to make assumptions in order to simplify the problem. Justify these assumptions and comment on how they may or may not affect the final result.
- The final report should be thoughtful, well-written and neatly organized. It should summarize your approach to the problem and present your conclusions with full explanation. The mathematical detail of your work should also be presented in some fashion. (Depending on the particular project, it might be part of the solution summary or included as an appendix to a less technical document.)
- If any questions persist (e.g. Have we investigated all aspects of the project?) or there is lack of clarity on some point (e.g. how much mathematical detail to include), be certain to discuss them with your instructor *before* writing the final report.

Syllabus for Calculus I

The intent of this syllabus is to concentrate on ideas rather than on manipulations, since manipulations can be more conveniently carried out with the aid of a hand-held calculator or a computer algebra system. After each topic, an estimated number of classes to cover the material is given in parentheses. These estimates were made based on a total of 32 class periods for the course. The 32 classes in the syllabus are intended to provide time for testing, applications, and enrichment topics as well as for the central ideas ordinarily covered in calculus. The projects from the first three sections of this volume are noted within the outline according to their prerequisite. However, we encourage use of the projects at other appropriate times as well. Annotated syllabi for Calculus I and II are contained in Volume I of this collection of resource materials.

CALCULUS I: THE DERIVATIVE AND THE INTEGRAL

1. Introduction (1)

2. Functions and Graphs (4)

   Cruise Control

3. The Derivative (10)

   Security System Design
   Crankshaft Design
   Valve Cover Design
   The Tape Deck Problem

4. Extreme Values (8)

   Optimal Design of a Steel Drum
   Finding the Most Economical Speed for Trucks
   Designer Polynomials
   Designing a Pipeline With Minimum Cost

5. Antiderivatives and Differential Equations (3)

   Population Growth
   Drug Dosage

6. The Definite Integral (6)

   Logarithms: You Figure It Out
   Numerical Integration and Error Estimation
   An Integral Existence Theorem
   A Fundamental Project
   Inventory Decisions
   Tile Design
   Minimizing the Area Between a Graph and Its Tangent Lines
   Riemann Sums, Integrals, and Average Values
   The Ice Cream Cone Problem

DERIVATIVES

The projects in this section require only knowledge of single-variable differential calculus. There are two primary concepts which these projects address. First, many of the projects involve using the derivative in some type of curve fitting application. This varies from a data fit ultimately aimed at constructing an optimization problem in *Finding the Most Economical Speed for Trucks*, to fitting data according to critical and inflection points in *Designer Polynomials*, to using data to construct models of technological equipment in *Cruise Control, Security System Design* and *The Tape Deck Problem*. Second, there are applications of single-variable optimization techniques to the "real world" in *Optimal Design of a Steel Drum, Designing a Pipeline With Minimum Cost, Crankshaft Design, Valve Cover Design* and *Finding the Most Economical Speed for Trucks*.

Title: Optimal Design of a Steel Drum

Author: John Ramsay, College of Wooster

Problem Statement: A 55-gallon Tight Head Steel Drum is constructed by attaching 18 gage (i.e. .0428 inches thick) steel disks to the top and bottom of a cylinder created by rolling up a 20 gage (i.e. .0324 inches thick) steel sheet.

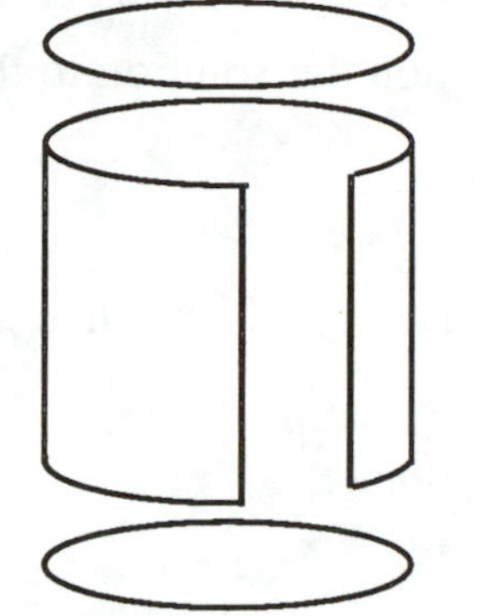

The vertical seam on the cylinder is welded together and the top and bottom are attached by a pressing/sealing machine. The pressing/sealing process requires approximately $\frac{13}{16}$ inches from the cylinder and $\frac{3}{4}$ inches from the disk to be curled together and hence these inches are lost in the final dimensions. In addition, the top and bottom are set down $\frac{5}{8}$ inches into the cylinder. For clarification, refer to the American National Standard (ANSI) specification diagram below.

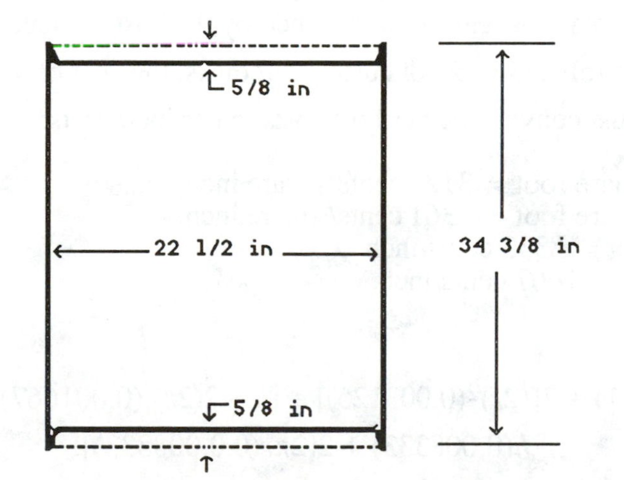

Steel can be purchased in coils (rolls) of any specified width. Construction costs can be summarized as follows:

18 gage steel is 45 cents/square foot
20 gage steel is 34 cents/square foot
welding and pressing/sealing cost 10 cents/foot
cutting steel costs 2 cents/foot.

Is the ANSI specified drum the most efficient use of material in order to obtain the required 57.20 gallon minimum volume capacity of a 55 gallon drum? Fully justify your answer.

Information for the instructor only:

Problem abstract: This project is designed to give students a more complete experience in applying calculus to a realistic problem. As a result of beginning with unpolished information, students must do almost everything in setting up the model, solving the problem and presenting their solution. While strengthening their understanding of mathematical concepts involved in extrema theory, it will bring the application of the material to life. The students must make decisions regarding the important aspects of the construction of the drum and determine relevant costs before setting up and solving the optimization model. Under most methods of construction the solution to the model will involve solving a cubic or quartic equation.

Prerequisite skills and knowledge: Theory of extrema.

Essential/useful library resources: none

Essential/useful computational resources: Under most methods of construction the solution to the model will involve solving a cubic or quartic equation and hence it will be necessary that they have access to equipment suited for solving such equations (graphically or numerically).

Example of an acceptable approach: If we let r represent the original disk radius in inches and h the original cylinder height, then 20 gage steel should be bought in coils of width h and 18 gage steel should be bought in coils of width $2r$. (Disks will be cut out of $2r$ by $2r$ squares, yielding some wasted material in construction.) Total cost can be determined by the cost of steel on sides, top and bottom, the cost of cutting pieces out of rolls, the cost of cutting out disks, the cost of welding the seam, and the cost of pressing/sealing. We must convert the per foot costs to per inch costs:

45 cents/square foot = .3125 cents/square inch
34 cents/square foot = .2361 cents/square inch
10 cents/foot = .8333 cents/inch
2 cents/foot = .1667 cents/inch.

Hence

$$\text{Cost} = 2\pi rh(0.002361) + 2[(2r)^2(0.003125)] + [h + 2(2r)](0.001667) + 2[2\pi r(0.001667)] + h(0.008333) + 2[2\pi r(0.0.008333)].$$

We use that 57.20 gallons = 13,213 in^3 (1 gal. = 231 in^3) and calculate the inner height and inner radius of the completed drum as follows:

Height $= h -$ sealing loss at top and bottom $-$ end indention loss at top and bottom

$= h - 2 \times \frac{13}{16}$ in. - $2 \times \frac{5}{8}$ in.

$= h - \frac{23}{8}$ in.

Radius $= r -$ sealing loss $-$ end indention loss

$= r - \frac{3}{4}$ in. $- \frac{5}{8}$ in. $= r - \frac{11}{8}$ in.

Thus we have the constraint $\pi (r - \frac{11}{8})^2 (h - \frac{23}{8}) = 13{,}213$.

Solving, one gets a minimum at approximately $r = 11\frac{15}{16}$ inches, $h = 40\frac{7}{16}$ inches with cost = \$12.73. The ANSI specified drum uses $r = 11\frac{1}{4} + \frac{11}{8} = 12\frac{5}{8}$ inches and $h = 34\frac{3}{8} + \frac{13}{8} = 36$ inches with cost = \$12.78. The above solution yields a savings of \$.05/drum.

Conjectures we expect that some students will make: Students will likely make a variety of assumptions as to how much material is "lost" in the construction process.

Questions for further exploration: Students who quickly produce a good solution to the problem could be encouraged to do sensitivity analysis on the cost function coefficients. Also, one can speculate as to why steel drums and food cans do not seem to be constructed with dimensions which yield minimal cost. For example, steel drums need to be easily portable and cans on shelves need to be visually pleasing (the Jolly Green Giant doesn't look so good on a short, fat can). Interested students could do a similar exercise with canned goods and speculate why particular shapes are chosen over the discovered "most economical" can. It is interesting to note how much closer bulk food containers fit the "correct" dimensions.

Special evaluation suggestions: One of the intentions of a problem such as this one is to have the students analyze and justify the assumptions they have made in constructing their particular model. A written or oral communication of this should accompany their solution.

Title: Finding the Most Economical Speed for Trucks

Author: John Ramsay, College of Wooster

Problem Statement: A trucking company would like to determine the highway speed that they should require of their drivers. The decision is to be made purely on economical grounds, and the two primary factors to be considered are driver wages and fuel consumption. Wage information is easily obtained: drivers earn from $11.00 to $15.00 an hour, depending on experience. Incorporating the fuel consumption question is much more difficult and the company has hired you as consultants in order to solve the problem for them. Correctly assuming that fuel consumption is closely related to fuel economy at various highway speeds, they have provided you with the following statistics taken from a U. S. Department of Transportation study*:

MILES PER GALLON AT SELECTED SPEEDS

VEHICLE	50 mph	55 mph	60 mph	65 mph
Truck #1	5.12	5.06	4.71	#
Truck #2	5.41	5.02	4.59	4.08
Truck #3	5.45	4.97	4.52	#
Truck #4	5.21	4.90	4.88	4.47
Truck #5	4.49	4.40	4.14	3.72
Truck #6	4.97	4.51	4.42	#

#Due to laws controlling fuel injection, this vehicle could not be operated at 65 mph.

The company expects a written report with your recommendation. The report should include justification for your conclusion.

Note: Though the report should include the mathematical detail of your work, the overall presentation should not assume the reader has had experience with the techniques you employ.

* U.S. Department of Transportation. *The Effect of Speed on Truck Fuel Consumption Rates*, by E. M. Cope. [Washington]: U. S. Department of Transportation, Federal Highway Administration, Office of Highway Planning, Highway Statistics Division. 1974. (TD2.2:Sp3)

Information for the instructor only:

Problem abstract: This project is interesting because it provides students with real data and asks them to solve a realistic problem. Working from very little initial information and obtaining a pleasing solution is very satisfying for students and will serve to boost confidence and help them see the relevance of the mathematics involved. There are many useful skills involved in the project: library research (possibly), data consolidation, curve fitting, optimization and communication of their solution.

Prerequisite skills and knowledge: Single variable theory of extrema. It would be very helpful if the students have already had some experience with curve fitting using either linear regression or systems of equations to develop polynomial functions which fit the given data. Students who have had no such experience will need some guidance in this.

Essential/useful library resources: The statistical data has been taken from a U.S. Department of Transportation Report prepared by the Federal Highway Administration. This document is a special government publication and the title is indicated on the problem page. If the library at your institution has this document, students could be directed to it in order to find the statistics rather than giving the data to them in the problem statement. You should check your library first since most libraries only carry partial government document collections.

Essential/useful computational resources: If polynomial fits are used, solving systems of equations with more than two or three unknowns is difficult without some sort of symbolic computation software. If regression is used, then some statistical software would be helpful.

Example of an acceptable approach: The function relation between mpg and mph can be modeled many ways. We include here a cubic polynomial fit and a linear regression fit.

1) To do a polynomial fit we need to obtain one mpg for each relevant mph. One way we can do this is by computing the average mpg at each speed. This yields the data points

$$(50, 5.11), (55, 4.81), (60, 4.54), (65, 4.09).$$

Using computer algebra software we can solve the system of four equations that results from plugging these four data points into the general cubic, $y = ax^3 + bx^2 + cx + d$. The resulting cubic is

$$f(x) = -0.00028x^3 + 0.0468x^2 - 2.657x + 55.96.$$

2) If we do a linear regression on all data points we obtain

$$f(x) = 8.3356 - 0.06414x.$$

In either case we now need to incorporate f into a cost function. Let D represent distance traveled (this turns out to be irrelevant), F represent cost of fuel per gallon, and W represent driver's hourly wage. Also note that x above represents speed. We obtain the following cost equation:

$$\text{Cost} = W\frac{D}{x} + F\frac{D}{f(x)}.$$

At the time of this writing, $F = \$1.00$ was reasonable. (Students should be expected to determine this value themselves.) Applying extrema theory to this with $F = 1$ and $W = 11, 12, 13, 14$, then 15 we get minimum cost at the speeds indicated in the following table:

Driver wage	Ideal Speed	
	Cubic Fit	Linear Fit
\$11.00	59.8 mph	59.3 mph
\$12.00	60.3 mph	60.7 mph
\$13.00	60.8 mph	62.0 mph
\$14.00	61.2 mph	63.2 mph
\$15.00	61.5 mph	64.4 mph

Conjectures we expect that some students will make: Students are likely to make any of a number of assumptions with regard to how to use the given data. For example, they may do the averaging as above or they may solve the problem for each truck and either report all results or report some form of average result. Also, they may decide to fix a specific driver wage rather than report on several wages.

Questions for further exploration: For lighter weight trucks, mpg. figures are much higher. Students could be asked to analyze the sensitivity of their solution to changes in data, particularly higher miles per gallon figures.

References/bibliography/related topics: There are some statistics on fuel efficiency of small trucks. The documents which contain this information are difficult to find but a government documents reference librarian should be able to help students find what they need.

Special implementation suggestions: There are several difficult steps in this project. First, if the students have not done or seen any curve fitting, it will take them some time to get comfortable with the concept. Second, the cost function in this problem is not easy. Determining how to convert natural variables (speed and driving time) to corresponding cost (driver total wages and total fuel costs) is difficult. Finally, the role of distance traveled is likely to be a point of confusion. For these reasons, an intermediate progress report is highly recommended.

Special evaluation suggestions: There are a wide variety of approaches to solving this problem. Students should be expected to justify the chosen approach and its conclusions.

Title: Designer Polynomials

Author: Charles Jones, Grinnell College

Problem Statement: For this project, imagine that you are a calculus instructor needing some "nice" examples of functions for your class. Your purpose is to help the students understand the relationships between polynomial functions and their graphs. In particular, you are interested in quadratic functions, $x^2 + cx + d$, cubic functions, $x^3 + bx^2 + cx + d$, and quartic functions, $x^4 + ax^3 + bx^2 + cx + d$, where a, b, c, and d all have integer values. Thoroughly answer the following questions, giving complete justification for each result. Your final report will be evaluated based on mathematical correctness; clarity of presentation; and correct use of the English language, including structure, grammar and spelling.

A. Quadratic functions of the form $f(x) = x^2 + cx + d$.

(1) How many critical points are there for each choice of c and d?

(2) Are the critical points maxima, minima or neither?

(3) How should you choose c and d to insure that all critical points occur at integer values of x?

(4) What must the shape of the graph of f be?

B. Cubic functions of the form $g(x) = x^3 + bx^2 + cx + d$.

(5) How many inflection points are there?

(6) Produce examples where the inflection points and critical points all occur at integer values of x where $g(x)$ has:

(a) 0 critical points,

(b) 1 critical point,

(c) 2 critical points,

(7) Write general rules for choosing b, c, and d to produce families of examples in question 6.

C. Quartic functions of the form $h(x) = x^4 + ax^3 + bx^2 + cx + d$.

(8) Choose a, b, c, and d so that $h(x)$ has 3 critical points at integer values of x. Produce some general techniques to generate a family of examples with 3 critical points, all at integer values of x. (Hint: Rather than solving cubic equations to determine if there are 3 integer-valued roots, start with the roots and produce the cubic equation. For example, given the roots 2, 3, and –1, the equation is $(x - 2)(x - 3)(x + 1) = 0$ or $x^3 - 4x^2 + x + 6 = 0$.)

Information for the instructor only:

Problem abstract: This project reverses the common textbook approach where the students are given a polynomial function, then asked to find and classify critical points, inflection points, and graph the function. The purpose of the project is two-fold: first, by working with functions and their derivatives from a new perspective, the students should acquire a better appreciation of the relationships between derivatives and properties of the graphs of functions. Second, this project should deepen each student's understanding of functions on a more basic level because the student does not have the opportunity to mechanically operate on a given formula for a function. Rather, the student is forced to first focus on required properties of a function and then do some creative work to produce a specific formula that satisfies those properties.

Prerequisite skills and knowledge: Students need algebra skills including knowledge of the relationship between roots and linear factors of a polynomial. This project would fit well in a calculus course after introductory work on finding critical points and inflection points using the derivative.

Essential/useful library resources: none

Essential/useful computational resources: None is essential; however, a graphing package could be useful.

Example of an acceptable approach:

(1) $f(x) = x^2 + cx + d; f''(x) = 2x + c$. So $f''(x) = 0$ if and only if $x = \frac{-c}{2}$. There is always one critical point for each choice of c and d; the choice of d does not affect the x coordinate of the critical point.

(2) The critical point must be a minimum. Several justifications are possible, for example, the second derivative test.

(3) To insure that the critical point occurs at an integer value of x, c must be even. The value of d is irrelevant.

(4) Its shape will always be a parabola opening upward.

(5) $g(x) = x^3 + bx^2 + cx + d$; $g'(x) = 3x^2 + 2bx + c$; $g''(x) = 6x + 2b$. So, the one inflection point always occurs where $x = \frac{-b}{3}$.

(6) All three parts of this question depend on $g'(x) = 3x^2 + 2bx + c = 0$.

(a) No solutions: $(2b)^2 - 4(3)(c) < 0$ must hold. This gives $c > \frac{b^2}{3}$. Further, from question (5), we know that b must be a multiple of 3 if the inflection point is to have an integer value for x. One particular example is $b = -6$, $c = 38$, $d = 100$, yielding the function given by $g(x) = x^3 - 6x^2 + 38x + 100$.

(b) One solution: $(2b)^2 - 4(3)(c) = 0$ must hold. This gives $c = \frac{b^2}{3}$. Again, we know that b must be a multiple of 3. The critical point of any satisfactory function occurs at $x = \frac{-(2b)}{2(3)} = \frac{-b}{3}$, the same x–coordinate as the inflection point. One particular example is $g(x) = x^3 - 6x^2 + 12x + 100$.

(c) Two solutions: $(2b)^2 - 4(3)(c) > 0$ must hold. This gives $c < \frac{b^2}{3}$. Now, to insure that the critical points occur at integer values of x, we need to think about the form of $g'(x) = 3x^2 + 2bx + c = 0$. Since b must be a multiple of 3, we can let $b = 3k$ and the formula becomes $3x^2 + 6kx + c = 0$. If we were able to factor a 3 out of this equation, we would be closer to the goal; so let $c = 3m$, and we get $3x^2 + 6kx + 3m = 0$ or $x^2 + 2kx + m = 0$. We could use either of two techniques to proceed from here; one would be to use the discriminant of the quadratic formula and choose k and m so that $(2k)^2 - 4m$ is a perfect square, say p^2. Then the two critical points would be $\frac{p-2k}{2}$ and $\frac{p+2k}{-2}$. For example, if $k = 5$ and $m = 9$, then $p = 8$ and the critical points are -1 and -9. If we choose $d = 50$, we get a good example: $g(x) = x^3 + 15x^2 + 27x + 50$. The other approach is to pick two roots and then determine k and m. For example, let 6 and -2 be the two roots. Then $(x - 6)(x + 2) = x^2 - 4x - 12 = 0$ yields $k = -2$, $m = -12$ and $b = -6$, $c = -36$. Then if we choose $d = 50$, we get a good example: $g(x) = x^3 - 6x^2 - 36x + 50$.

(7) The answers to this question are open-ended. For (a), one correct answer is to let $b = 3k$ (k any integer), $c = 3k^2 + 1$, and $d =$ any integer. For (b), one correct answer is to let $b = 3k$, $c = 3k^2$, and d be any integer. For (c), either of the techniques described in (6c) would be a good reply.

(8) $h(x) = x^4 + ax^3 + bx^2 + cx + d$; $h'(x) = 4x^3 + 3ax^2 + 2bx + c$. If we choose to let $a = 4k$, $b = 2m$, and $c = 4n$ (k, m, n integers), then $h'(x) = 0$ becomes $4x^3 + 12kx^2 + 4mx + 4n = x^3 + 3kx^2 + mx + n = 0$. As mentioned in the hint, a general technique is to pick 3 roots at integer values, then determine k, m, n, and hence, a, b, and c. The 3 roots picked will have to add to a multiple of 3 because the coefficient of x^2 is $3k$. For example, let the roots be -1, 2, and 5. Then

$$(x + 1)(x - 2)(x - 5) = x^3 - 6x^2 + 3x + 10 = 0.$$

Thus, $k = -2$, $m = 3$, $n = 10$, and $a = -8$, $b = 6$, and $c = 40$. So a good example of a quartic function with 3 critical points at integer values of x is $h(x) = x^4 - 8x^3 + 6x^2 + 40x - 15$.

Conjectures we expect that some students will make: Coming up with "nice" examples is not as simple as one might first think.

Questions for further exploration:

(9) Can you find an example with 3 critical points and 2 inflections points, where all critical and inflection points occur at integer values of x? (This is a very difficult problem, you will do very

well to make any progress on it. The following solution involved a few pages of algebra followed by using a computer program to search for combinations of critical points and inflections points that satisfied the algebraic requirements. This is the simplest example I know of: $h(x) = x^4 - 44x^3 + 432x^2$, which has critical points at $x = 0$, 9, and 24; and inflection points at $x = 4$ and 18. I would be very pleased to hear of any examples you or your students may find.)

(10) For the quartic, $h(x) = x^4 + ax^3 + bx^2 + cx + d$, what possibilities are there with regard to the number of critical points and the number of inflection points? Investigate, give examples, and produce general rules where possible. For example, if there are two inflection points and two critical points, must one of the critical points also be an inflection point? One example of this is $h(x) = x^4 + 4x^3 + 1$, which has critical points at $x = -3$ and 0, and inflection points at -2 and 0.

Title: Cruise Control

Authors: Eric Robinson, John Maceli, Diane Schwartz,
Stan Seltzer, and Steve Hilbert, Ithaca College

Problem Statement: You have been hired as a mathematical analyst for a major car manufacturer. A "cruise control" system is to be designed for a mid-size car. The problem is first broken down into two parts: designing a system to convert real speed into recorded speed; and designing a mechanical apparatus which either slows down or speeds up the car depending upon its recorded speed and a "speed set." Your job is to solve the first part of the problem.

The solution to the problem of converting real speed into recorded speed is again split into two stages. In the first stage, a metal pin is secured to the inside of the hub of the right front wheel of the car. This pin registers one unit on a counter with each revolution of the wheel. (See Figure 1.)

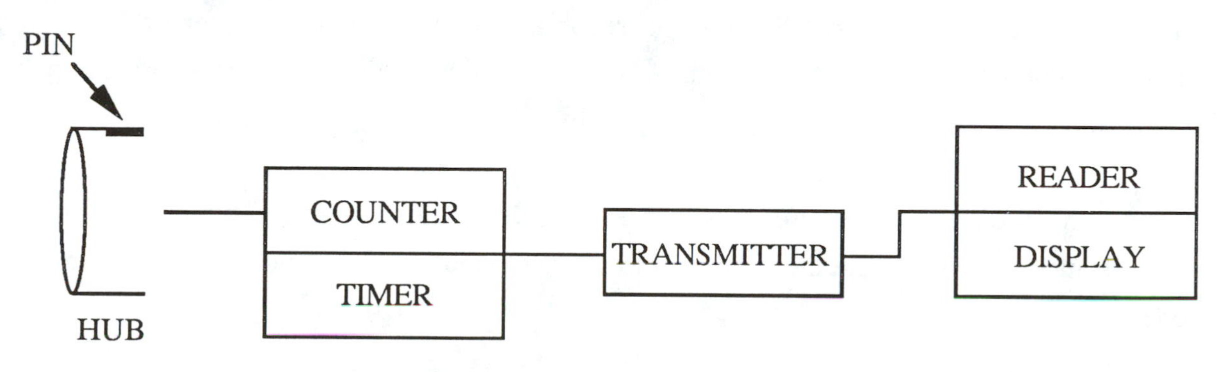

Figure 1

The counter is connected to a timer and a transmitter. In the second stage, the transmitter transmits the total count (i.e. number of revolutions of the wheel) to a "reader" each second (hence the need for the timer) and then the counter is cleared (i.e. set back to zero). The reader converts each such count into a recorded speed and displays that speed on a digital speedometer.

The hub of the wheel has a radius of seven inches. Furthermore, the distance from the center of the wheel to the ground is twelve inches.

A) Give a rule for a function which has as its input (domain) the real speed of the car and which has as its output (range) the count that would appear on the counter each second for a car traveling at that speed. Draw the graph of this function. Limit the domain of your graph to the interval [50, 60].

B) Give a rule for a function which has as its input the count per second that could be transmitted by the transmitter and which has as its output the recorded speed of the car. Draw the graph of this function.

C) Explain how the functions in A) and B) should be combined to give a function that has as its input the real speed of the car and as its output the recorded speed of the car. Carefully, draw the graph of this function over the interval [50, 60].

D) Assume now that the cruise control device is about to be constructed. You, however, can see from examining your results thus far that there might be a problem when the control is "set" at 55 mph. Explain.

E) Two design alterations have been suggested to make the system better. One suggestion is to add additional equally spaced pins around the hub of the wheel. (Say, for example, there would be 4 pins around the hub.) The formula the reader uses to convert the transmitted count into a recorded speed would have to be changed, of course. The second suggestion would be to transmit the count every 0.5 seconds instead of every second. How would the graphs in A), B) and C) change with either of these suggestions? Would you recommend implementing either or both of these suggestions? Why? Are there other improvements that you would suggest making? Support your suggestions.

F) Suppose that the car manufacturer now wishes to devise an instrument that could keep track of the total distance traveled while the cruise control is activated. How would you suggest that this be done?

Information for the instructor only:

Problem abstract: This project is designed for early use in Calculus I. It serves as an application of the concepts of composition of functions, piecewise - defined functions, and graphical analysis of functions. The project also serves to motivate the concepts of limits and continuity. Most importantly, the project forces the students to deal with a function as a device which determines a unique output for a given input, rather than as just an algebraic formula.

Prerequisite skills and knowledge: Functions as represented by graphs and formulas, composition of functions

Essential/useful library resources: none

Essential/useful computational resources: none

Example of an acceptable approach: The following table lists the correspondence between clicks/second (that's 24π inches/second) and miles per hour.

Clicks/second	MPH
1	4.284
11	47.124
12	51.408
13	55.692
14	59.976
15	64.260

A) Using the above table, one can derive a formula using a piecewise definition or the greatest integer (or least integer) function. The students should graph this step function.

B) Again, the above table can be used to give a piecewise definition or one could give a formula such as

$$f(C) = \text{Round}\ (4.2840 * C)$$

where C is the clicks/sec and Round denotes the rounding to the nearest integer function.

C) Composing the answers to A and B yield another step function, not the identity. For example, if the real speed of 54 mph is the input, the output will be 51 or 56 mph, depending on decisions the student makes.

D) As discussed above, since 55 mph does not correspond to an integer number of clicks/sec., the recorded output will not be 55 mph, regardless of the actual speed.

E) Increasing the number of pins causes uniform, but not monotonic, convergence of recorded speed to actual speed (up to the precision of the recorded speed).

Shortening the transmission period will lessen the accuracy.

F) One suggestion is to use a counter that does not get reset every second.

Title: Security System Design

Authors: Steve Hilbert, John Maceli, Eric Robinson,
Diane Schwartz, and Stan Seltzer, Ithaca College

Problem Statement: You are designing a security system for a hospital. The hospital keeps its supply of drugs in a storeroom whose entrance is located in the middle of a 40 foot long hallway. The entrance is a three foot wide door. The hospital wishes to monitor the entire hallway as well as the storeroom door. You must decide how to program a detector to accomplish this. The detector runs on a track and points a beam of light straight ahead on the opposite wall. The beam reaches from the floor to the ceiling. Think of the hallway as a coordinate line with the middle of the door at the origin and the hallway to be watched as the interval [–20, 20]. You need to decide what $x(t)$ is for t, where $x(t)$ represents the position of the beam at time t (in seconds).

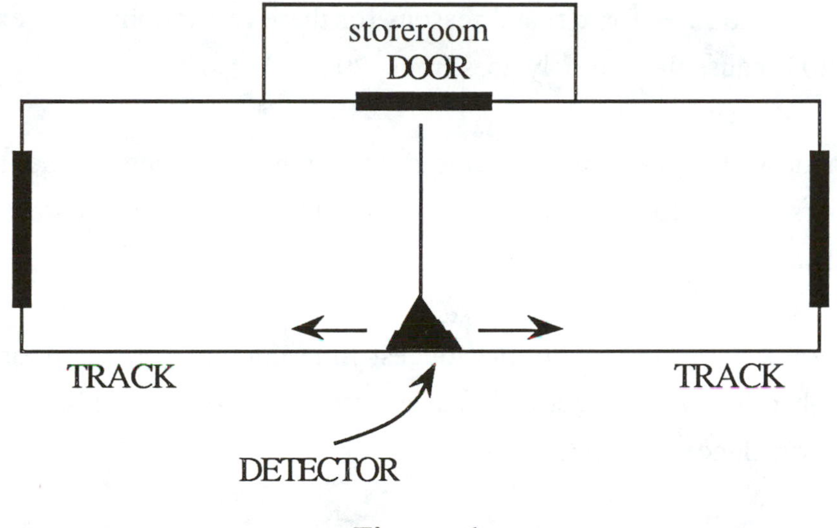

Figure 1

Figure 1 shows the beam pointing at the origin (i.e. the middle of the door, so if the detector was at this position at some time T we would write $x(T)=0$). As another example $x(5) = -15$ would mean that the beam is pointing at the part of the wall 15 feet to the left of the middle of the door 5 seconds after the detector starts.

Part 1:

a) Draw a graph of x versus time for what you think is a good choice for $x(t)$. Assume that t ranges from 0 to 10 minutes. Be sure to elaborate on why you think this is a good choice.

b) The beam must stay on an object for at least one-tenth of a second in order to detect that object. If the width of a person is 1 foot decide whether your answer to a) will detect a person standing anywhere in the hallway. Explain.

c) For your answer to a) compute the longest time interval that the door will not be under surveillance. Remember the door is 3 feet wide and assume that as long as the beam is hitting any part of the door it is under surveillance.

d) Investigate whether an intruder could get to the door by walking down the hallway without being detected by your system. Explain how she could do it and how likely you think it is. This may inspire you to revise your answer to a).

e) What if the intruder is running?

Part 2:

a) Find a rule (function) for $x(t)$ for the first 10 minutes. This part of your report should include any restrictions on possible rules for $x(t)$ and reasons for these restrictions. For example, $x(t)$ should never be less than -20 because the hall only goes from -20 to 20.

b) The beam must stay on an object for at least one-tenth of a second in order to detect that object. If the width of a person is 1 foot decide whether your answer to a) will detect a person standing anywhere in the hallway. Explain.

c) For your answer to a) compute the longest time interval that the door will not be under surveillance. Remember the door is 3 feet wide and assume that as long as the beam is hitting any part of the door it is under surveillance.

d) Investigate whether an intruder could get to the door by walking down the hallway without being detected by your system. Explain how she could do it and how likely you think it is. This may inspire you to revise your answer to a).

e) What if the intruder is running?

Part 3:

Relate your answers from parts 1 and 2.

Information for the instructor only:

Problem abstract: This is an open-ended project. The fact that there is no "right" solution is very enlightening for many students who think that problem-solving consists of searching for a similar problem that is worked out in the text.

This is a cyclical project. The students are asked to rethink their solution several times (e.g., parts 1a, 1b, 1d). They are also asked to approach the problem first on a geometrical level (part 1) and then on a computational level (part 2). Hopefully, this project will cause the students to realize the power of approaching a problem using both geometrical and computational techniques.

This project involves several mathematical concepts (depending on the students' approach): the relationship between position and velocity, amplitude and period of functions (probably sine or cosine), and sawtooth functions.

Prerequisite skills and knowledge: Derivatives, velocity as slope of the position function

Essential/useful library resources: none

Essential/useful computational resources: none

Example of an acceptable approach:

Part 1: One answer would be a sawtooth wave with amplitude 20 and slope 10 or –10, where the independent variable is time in seconds and the dependent variable is position in feet.

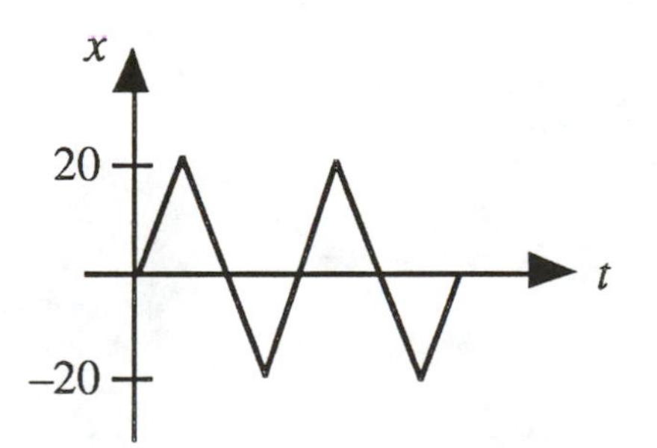

This scheme can always detect a person standing still under the assumptions in b). You may want the students to make the corners rounded since an instantaneous change in velocity from –10 to 10 is unrealistic; and if it could be done it would also be very hard on the detector.

There are obviously many other possible schemes. One type of alternative scheme is to have the detector scan from the storeroom door to the end of the hallway at 10 ft/sec (or slower if a moving person is to be detected) and then return to the storeroom door very rapidly (say 40 ft/sec) and spend a few seconds scanning the storeroom door (since protecting the storeroom is our ultimate goal). Other variations might include some sort of random behavior so the path of the detector would not be predictable to an intruder.

Parts 1d, 2d, and 4 are interesting since a person walking or running in the direction opposite the direction of the scanner will not stay in the beam very long and so may not be detected (obviously all of this depends on assumptions made concerning the speeds of the detector and the intruder).

Part 2: The students may opt to find a formula for their part 1 answer, but unless they used a sawtooth or sine wave, the formula may be elusive. Another approach would be to use a sine function for its computational simplicity, but it suffers from the disadvantage of moving most quickly at the storeroom door (minimizing the protection of the storeroom) and lingering at either end of the hallway. The students should discuss this type of disadvantage if they choose such a formula.

Part 3: This part could be very interesting if the students have an elaborate graphical solution to part 1 and a simple formula solution to part 2. Their pondering of this part should cause an appreciation of approximations, piecewise-defined functions, and mathematical modeling in general. This part may also be an excellent precursor of Fourier series.

Questions for further exploration:

1. For your answers to parts 1 and 2, describe how you could defeat your system given your complete knowledge of the system. Be sure to supply details about how you would traverse the hallway and the amount of time you spend to unlock, open and close the storeroom door.

2. Discuss how likely you think it would be for someone without knowledge of your system to be able to defeat it.

Title: Designing a Pipeline With Minimum Cost

Author: John Ramsay, College of Wooster

Problem Statement: A common problem encountered by the oil industry is determining the most cost effective pipeline route in connecting various wells in an oil fertile area. The attached map is a section of a U. S. Geographical Survey Contour Map of northeast Ohio with wetland(swamp) area outlined for clarity. An existing oil well is located approximately at the point labeled *B*. If a new well is to be dug at point *A*, a pipeline installation company must be directed as to where to lay connecting pipe. In consultation with the installation company, the following information has been obtained:

- Straight, two-inch coated pipe must be used at a cost of $1.50/ foot.
- A maximum of two elbow joints may be used. Assume that the elbow joints may be fabricated with any angle measure.
- In crossing normal terrain, installation cost is $1.20/ foot.
- Installation in the wetland area requires an additional Track Hoe at a cost of $60/ hour.
- In a 10 hour day, a Track Hoe can dig approximately 300 feet of trench.

Determine the pipeline route connecting the new well at *A* to the well at *B* which incurs the least cost.

Suggestions: First, solve the problem as if the wetland separating *A* and *B* were a rectangle, then improve on this solution by modeling the wetland area more accurately. Also, reduce the number of paths to consider before you begin modeling. For example, one need not consider a path around the swamp to the north since it is further than the path around the swamp to the south and both traverse only normal terrain.

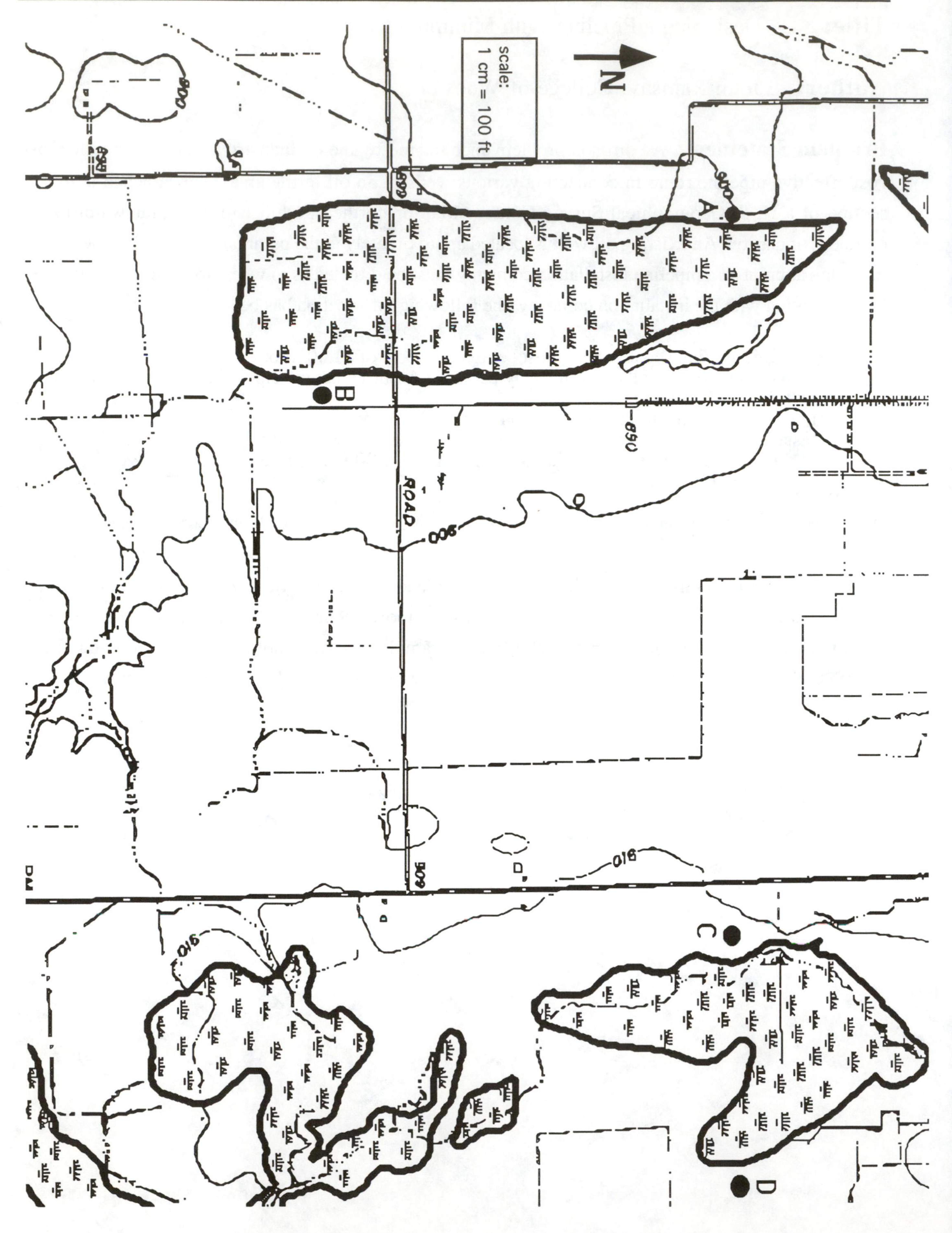
scale:
1 cm = 100 ft
N
A
B
C
D
ROAD

Information for the instructor only:

Problem abstract: This project is a more realistic version of a common textbook problem. It is surprising how quickly such a problem becomes extremely difficult. If the wetland in question cannot be approximated with a fairly simple polygon, a reasonable cost equation is hard to find. In this particular case, a rectangular or general trapezoidal shape can be used and both yield interesting results. Students will need to address the complexity of modeling a seemingly simple problem and should learn a great deal about making assumptions, sifting through given information and considering the many cases which must be dealt with in applied problems. We note that this becomes an excellent two-variable optimization problem if the elbow joint restriction is lifted.

Prerequisite skills and knowledge: Theory of extrema. If the two elbow joint restriction is lifted, the problem becomes a two-variable optimization.

Essential/useful library resources: none

Essential/useful computational resources: The solution will probably involve solving a system of equations involving square roots, which quickly reduce to quadratics and hence a calculator would be sufficient. However, a graphical or symbolic computation tool would be useful.

Example of an acceptable approach: Following the suggestion, we first model the swamp with a rectangle.

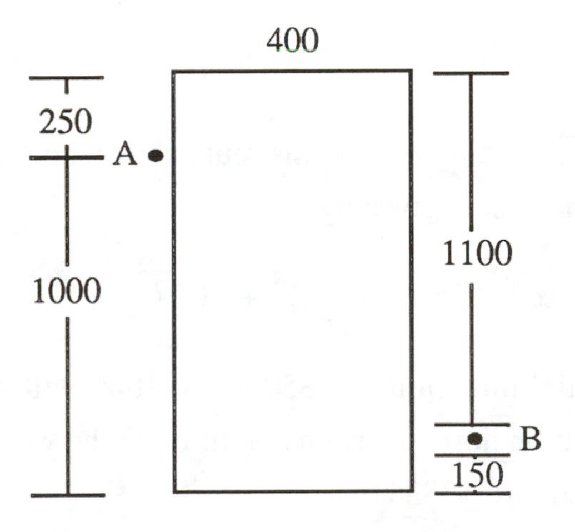

Clearly there is no point in moving west from point A. Cutting through the swamp in a northeast direction is also clearly not least expensive. Finally, laying pipe around the swamp to the north is further than laying pipe around the swamp to the south. Hence, from point A, we must move south along the swamp or begin cutting through the swamp in a southeasterly direction. We make the following notes for future reference:

Note 1: The cost of laying pipe around the swamp to the south is $(1.50 + 1.20) \times (1000 + 400 + 150) =$ \$4185.00.

Note 2: The cost of laying a straight line of pipe from A to B is $(1.50 + 1.20 + 2.00) \times \sqrt{850^2 + 400^2}$ = \$4415.25. (The additional \$2.00 in the cost per foot comes from $\dfrac{\$60\text{/hour additional track hoe}}{30\text{ feet/hour}} =$ \$2.00/foot.)

There are two reasonable models depending on whether we exit the swamp on the south side or the east side. If we let x denote the distance along the west bank before cutting across the swamp and y denote the distance covered along the exiting bank after exiting the swamp we can see the two models as follows:

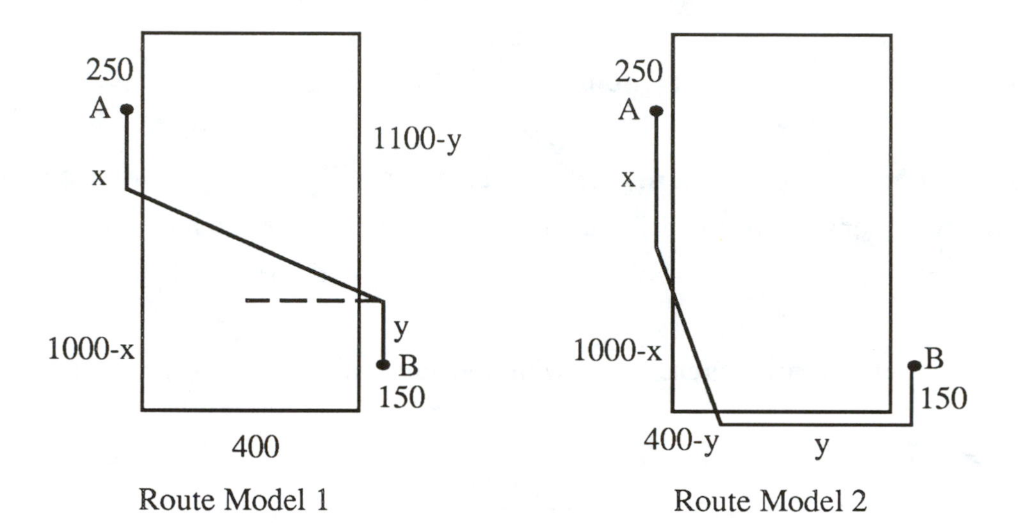

Route Model 1   Route Model 2

Route Model 1: A quick look at the diagram for this route shown above lets us assume, without loss of generality, that $y = 0$. Hence, the cost is given by

$$\text{cost} = 2.7x + 4.7\sqrt{(850 - x)^2 + 400^2} \qquad 0 \le x \le 850.$$

($x < 0$ implies a northward initial direction and $850 < x < 1000$ will always be more expensive than $x = 850$ as more pipe will be laid on normal terrain *and* through the wetland. The minimum in this case occurs at $x = 569$ with associated cost \$3834.

Route Model 2: Unless $y = 0$ or $y = 400$, this route violates the two elbow joint restriction. For any x, $y = 0$ yields a route clearly more expensive than the route of Note 2 above and $y = 400$ forces x to be 1000 and we have the route of Note 1 above. Hence this model contributes no new information. However, a full solution to the two variable optimization problem is included here in case the instructor chooses to lift the two elbow joint restriction.

The cost in this case is $2.7x + 4.7\sqrt{(1000-x)^2 + (400-y)^2} + 2.7y + (2.7)(150)$. This has a critical point at $x = 1000$, $y = 400$ (partials do not exist). There are no critical points inside the region $0 \le x \le 1000$, $0 \le y \le 400$. One can show that when $x = 0$ the cost is minimal for $y = 0$ and when $y = 0$ the cost is minimal when $x \approx 719$.

One must consider the boundary of the domain region. If $x = 1000$, we minimize the resulting function of y to get $y = 400$. Similarly, if $y = 400$ we minimize to get $x = 1000$. This is simply the route of Note 1 above and the associated cost is \$4185. If $x = 0$ the cost achieves its minimum value of \$5467 when $y = 0$. On the other hand, if $y = 0$ the cost is minimal for $x = 719$. The associated cost in this case if \$4644. We conclude that Route Model 2 has minimal cost \$4185 when $x = 1000$, $y = 400$. That is, we avoid the swamp entirely.

Thus, in the rectangular model we should go south from *A* 569 feet, then angle directly to *B* for a total cost of \$3834.

If we model the swamp more accurately, the problem begins to get even more complicated. Take the following model, for example:

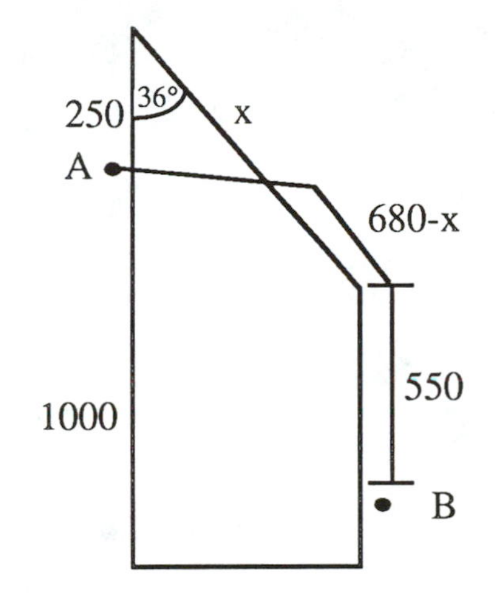

In this case, we obtain

$$\text{cost} = 2.7(550 + 680 - x) + 4.7\sqrt{x^2 + 250^2 - 2x(250)\cos(36°)}$$

which has a minimum at $x = 305$. This yields a cost of \$3340, the least found so far! Note that the solution to the rectangle model above covers the other set of possible paths relevant to this model. Also note that, as before, there is a discontinuity at $x = 0$. When we compute the cost of going around the swamp in this direction we get $2.7(250 + 680 + 550) = \$3996$, clearly not the minimum.

Conjectures we expect that some students will make: It is not likely that students will consider the discontinuity question addressed above. One would like, however, for them to see that the path around the swamp needs to be calculated separately in many cases.

Questions for further exploration: The same problem, using points C and D is a much more challenging one. It is quite open ended as far as what might be a "best" path. Students will need to make a number of assumptions in order to get at this one.

Special implementation suggestions: If students try to model the swamp with non-polygonal shapes, they will have great difficulty. They should either be told this at the outset or their progress should be monitored to keep them from running into this problem.

Title: Crankshaft Design

Author: Steve Boyce, Berea College

Problem Statement: In a reciprocating internal combustion engine, each piston is housed in a cylinder and attached to the rim of the crankshaft by a connecting rod as indicated in the figure below. The piston moves back and forth in the cylinder, and, in response, the crankshaft rotates.

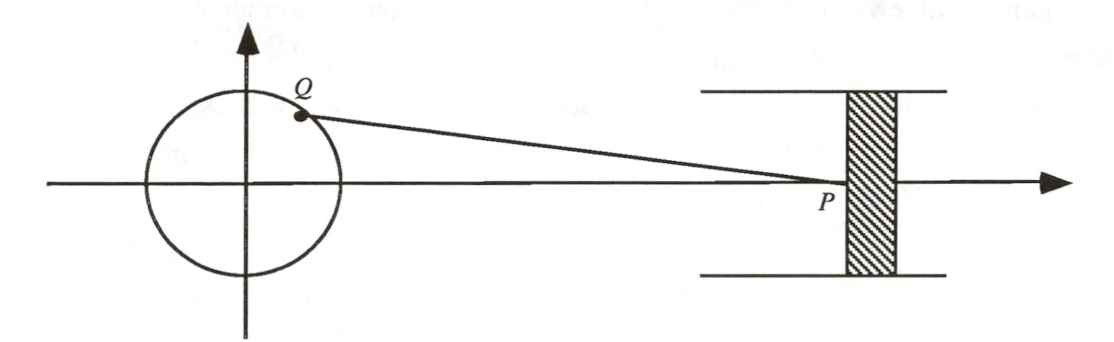

Among the factors which determine the stress on certain engine parts are the speed and acceleration of the pistons. This seems plausible in the case of a connecting rod, for example, since force is proportional to acceleration and one of the main forces exerted on a connecting rod comes directly from its linkage to the piston. One common indication of this relationship between stress and piston motion is the warning "red line" found on tachometers in some sports and racing cars. A tachometer displays engine speed measured in revolutions per minute (rpm's) of the crankshaft. To push an engine past its "red line" rpm level is to risk serious damage due to excessive stress on pistons, connecting rods and the linkages between the connecting rods and the pistons and the crankshaft. In this problem you are asked to investigate various aspects of the relationship between crankshaft rpm's, piston speed and acceleration, connecting rod length and crankshaft radius.

1) In a certain automobile, suppose a 60 miles per hour cruising speed results from the crankshaft rotating counterclockwise at the constant rate of 3000 rpm's. If the radius of the crankshaft is 1 inch and the length of the connecting rod is 4 inches, find the piston's maximum and minimum velocity (in feet/second accurate to the nearest hundredth) and acceleration (in feet /sec^2 accurate to the nearest whole number).

2) Is the piston motion in part (a) sinusoidal? That is, can the motion be described by a function of the form $x(t) = A + B\sin(Ct + D)$ for appropriately chosen constants A, B, C and D? [$x(t)$ is the x coordinate of P in feet after t seconds.]

3) a) In designing an engine, a decision must be made as to how long the connecting rods should be. Is it better to have them as short as possible or longer? How much longer?

   b) One question that arises in this connection is how changing the connecting rod length would affect piston velocity and acceleration. Assuming the crankshaft's rotational velocity and radius hold constant at 3000 rpm's and 1 inch respectively, investigate the relationship between connecting rod length and maximum absolute value of piston velocity and acceleration.

c) Do your results suggest any conclusions regarding ideal connecting rod length?

d) What other factors seem likely to have an important bearing on the question of connecting rod length?

4) Another design question concerns the radius of the crankshaft and what's gained or lost as it changes size. Assuming the crankshaft speed is 3000 rpm's and the connecting rod length is 4 inches, find the maximum absolute value of piston velocity and acceleration if the crankshaft radius is doubled to 2 inches. If you assume the engine can generate approximately the same average piston speed regardless of the crankshaft radius, what trade-off does your result suggest is involved in making the crankshaft radius larger? That is, would the larger radius seem more appropriate for a dump truck or a race car? Why?

Information for the instructor only:

Problem abstract: The goals of this project are to involve students in (a) geometric and trigonometric modeling; (b) the use of a computer and numerical estimation in performing max/min calculations that would be difficult to carry out analytically; (c) the use of a computer in investigating the effect of parameter variation; (d) the real-world interpretation of modeling results; (e) the clear and carefully reasoned written/oral explanation of analysis, conjecture and speculation related to an open-ended problem. While the problem statement does not pretend to be representative of the complexity involved in engine design, it is hoped that engaging it might at least tickle the student imagination--even for those heretofore totally uninformed about engines--regarding some possible interfaces between design questions and calculus.

Prerequisite skills and knowledge: This project can be assigned any time after students have been introduced to the first and second derivatives as velocity and acceleration and to the use of a software package capable of generating 2-dimensional graphs and enabling one to estimate with reasonable accuracy the maximum and minimum values of the first and second derivatives of a given function. However, without some additional knowledge (for example, familiarity with parametric equations) most students will find this project very challenging. Analytic work can lead to some nice insight, especially in part (3). For that purpose, it would be necessary to obtain (either by hand or through use of a computer algebra system) a derivative formula requiring the differentiation of sine and cosine functions and use of the chain rule. Reaching an appropriate conclusion in part (4) would be facilitated by some acquaintance with the concept of torque, at least to the extent of understanding that the strength of the turning effect on the crankshaft increases as the radius of the shaft increases.

Essential/useful library resources: none

Essential/useful computational resources: It is essential to have a software package which can be used to estimate with reasonable accuracy the maximum and minimum values of the first and second derivatives of a given function. It would be useful but not essential to have software capable of generating 2-dimensional graphs and obtaining simplified formulas for derivatives.

Example of an acceptable approach:

1) The stumbling block in part (1) may well be finding an appropriate function model for the piston's motion. To do so, first note that t seconds after beginning from (1,0) the point Q in Figure 1 has coordinates $(\cos(100\pi t),\sin(100\pi t))$ since 3000 revolutions per minute is equivalent to one revolution every 1/50 of a second.

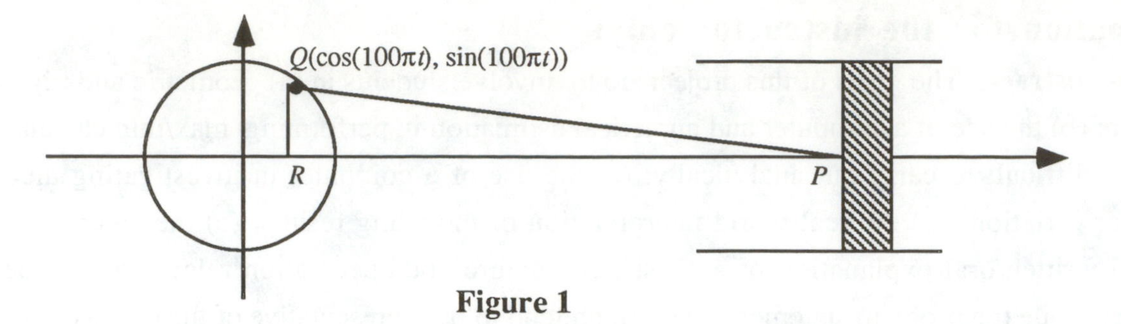

Figure 1

It follows that $x(t)$, the x coordinate of P in feet after t seconds, is given by

$$x(t) = (\cos(100\pi t) + \text{the length of } RP)\ /\ 12$$

$$= \frac{1}{12}(\cos(100\pi t) + \sqrt{16 - \sin(100\pi t)^2}\ ) \qquad \text{(by the Pythagorean Theorem).}$$

As one would expect, a graph of the first two periods of $x(t)$, shown is Figure 2, has a sinusoidal appearance.

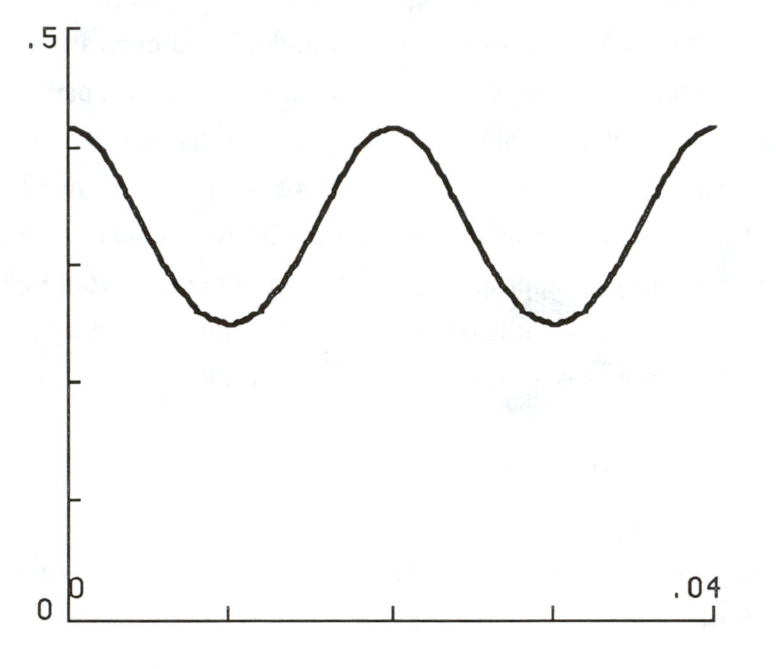

Figure 2

While the rest of part (1) is straightforward in concept, hand calculation of the derivatives and their zeroes would be a daunting task. It is a good time to use an available software package to obtain estimates of the maxima and minima of the velocity and acceleration. One-period velocity and acceleration graphs and extrema estimates are given below.

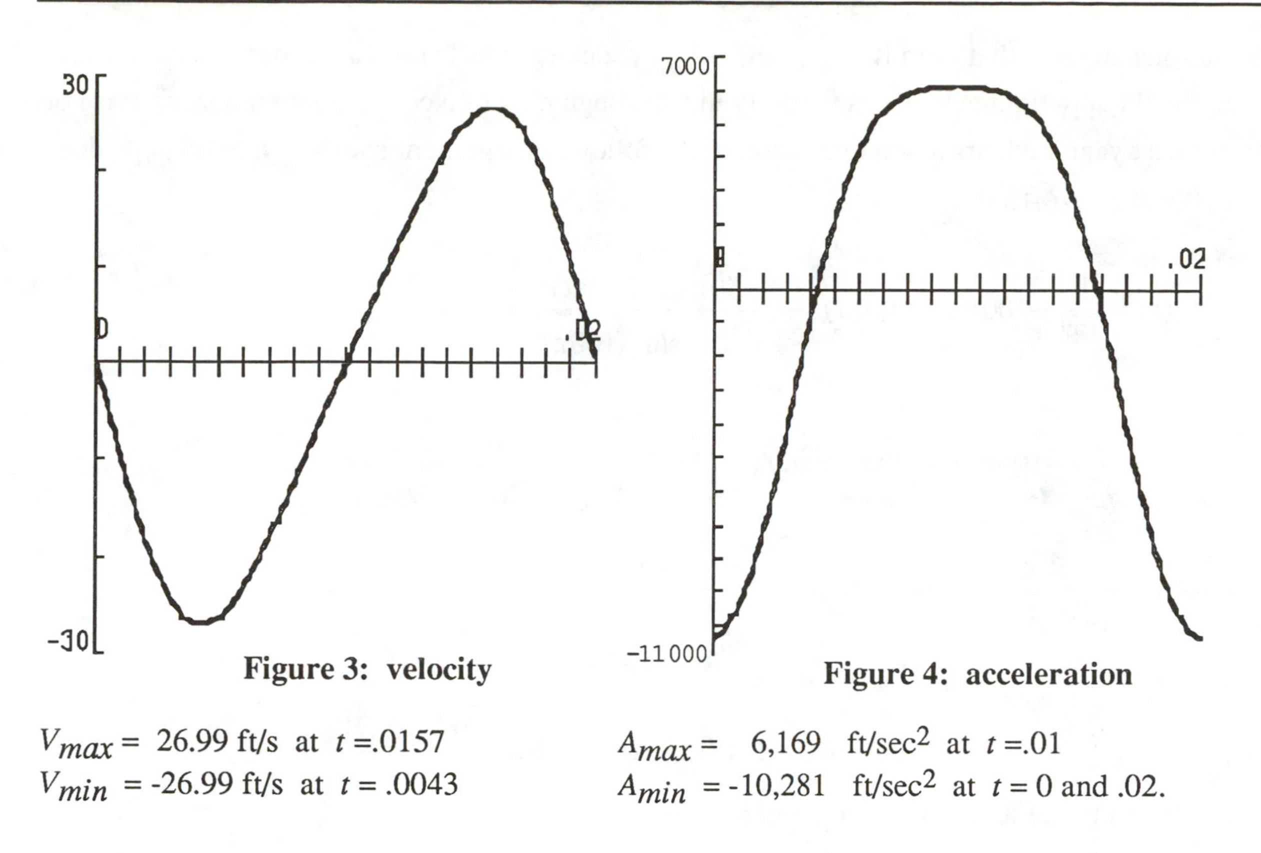

Figure 3: velocity

Figure 4: acceleration

V_{max} = 26.99 ft/s at t =.0157
V_{min} = -26.99 ft/s at t = .0043

A_{max} = 6,169 ft/sec^2 at t =.01
A_{min} = -10,281 ft/sec^2 at t = 0 and .02.

2) The appearance of the graphs in Figures 2 and 3 and the fact that $x(t)$ is generated by uniform circular motion suggest that the piston's motion is sinusoidal. So it is interesting to note that this is not actually the case. This is apparent from the graph of the second derivative which indicates that the curvature of x's graph is greater at its relative minima than at its relative maxima. Alternatively students might note that if x were sinusoidal then x'' would be also since $x(t) = A + B\sin(Ct + D)$ implies that $x''(t) = -C^2B \sin(Ct + D)$; it is clear from the lack of symmetry about the x-axis that x'' does not have this form. Another explanation might be based on a table of values for x and the observation that small increments in t cause greater changes in $x(t)$ near 0 and .02 than they do near .01.

3) The most likely approach is to estimate, with computer assistance, maximum absolute values for the velocity and acceleration of

$$x(t) = \frac{1}{12}\left(\cos(100\pi t) + \sqrt{L^2 - \sin^2(100\pi t)}\right)$$

for various values of the connecting rod length L. The following table contains such estimates for integer values of L from 2 (the smallest possible L value) to 7 along with the first period times when they occur.

L	$\lvert V\rvert_{max}$ f/s	t	$\lvert A\rvert_{max}$ f/s^2	t
2	29.41	.00376, .01624	12,337	0, .02
3	27.61	.00407, .01593	10,966	0, .02
4	26.99	.00430, .0157	10,281	0, .02
5	26.7	.00440, .0156	9,870	0, .02
6	26.54	.00450, .0155	9,595	0, .02
7	26.45	.00456, .01544	9,399	0, .02

These calculations suggest that both $|V|_{max}$ and $|A|_{max}$ decrease as L increases, but at a decreasing rate. In fact, the $|V|_{max}$ figures look suspiciously like the beginning of a convergent sequence. It is not beyond strong first year students to uncover parts of the following argument showing that $|V|_{max}$ does converge to $100\pi/12 \approx 26.18$ ft/s:

$$x'(t) = \frac{1}{12}\left(-100\pi\sin(100\pi t) - \frac{50\pi\sin(200\pi t)}{\sqrt{L^2 - \sin^2(100\pi t)}}\right) = s(t) + f(t, L) \text{ where}$$

$$s(t) = \frac{1}{12}\cdot -100\pi\sin(100\pi t) \text{ and } f(t,\ L) = -\frac{50\pi\sin(200\pi t)}{12\sqrt{L^2 - \sin^2(100\pi t)}}.$$

Note that $|f(t, L)| \le \dfrac{50\pi}{12\sqrt{L^2 - 1}} = M(L)$.

It follows that

$$s(t) - M(L) \le x'(t) \le s(t) + M(L),$$

and taking the maximum of all parts with respect to t yields

$$\max\ (s(t)) - M(L) \le \max\ x'(t) \le \max\ (s(t)) + M(L).$$

Since

$$\lim_{L\to\infty} M(L) = 0, \text{ we have } \lim_{L\to\infty}\ \max x'(t) = \lim_{L\to\infty}\ s(t) = -100\pi/12 \approx 26.18.$$

A similar (but messier) argument can be used to show that

$$\lim_{L\to\infty}\ \max x''(t) = \lim_{L\to\infty}\ s'(t) = (100\pi)^2 / 12 \approx 8225.$$

With or without the convergence arguments, the results of numerical experimentation (see, for example, those reported in the table above) are enough to suggest that (1) if piston velocity and acceleration were the only considerations, then the longer the connecting rods the better; (2) velocity and acceleration are most sensitive to changes in L when L is near its minimum value of 2; and (3) after a while, further increases in L yield only very small reductions in piston velocity and acceleration. Examples of other factors which would influence any decision about connecting rod length include engine block size and weight (the longer the connecting rods, the bigger and heavier the block must be) and strength of the connecting rods (the longer the connecting rods, the greater is their tendency to flex under a load).

When the crankshaft radius doubles from 1 to 2 inches while all other factors are held constant, the piston velocity and acceleration more than double:

radius	L	$\lvert V\rvert_{max}$ f/s	$\lvert A\rvert_{max}$ f/s^2
1	4	26.99	10,281
2	4	58.8	24,674

4) Looked at another way this means that if the engine is able to generate about the same average piston speed, then the crankshaft rotational speed will diminish as the radius increases. Since it is crankshaft rotation that is translated by the transmission and differential into movement in automobiles and trucks, diminished rotational speed translates into diminished vehicular speed--all other factors being equal!

If speed is lost as crankshaft radius increases, what is gained? The answer is the torque since the crankshaft radius is the approximate length of the arm on which the force applied through the connecting rods acts. A reasonable expectation based on this analysis is that truck crankshafts would in general have larger radii than their counterparts in race cars.

Title: Valve Cover Design

Author: Steve Boyce, Berea College

Problem Statement: A device similar to the one pictured below is sometimes used within a larger assembly of mechanical parts to periodically cover and uncover valve openings. As the circular shaft turns counterclockwise, a rigid connecting rod QP causes the valve cover to move back and forth between A and B.

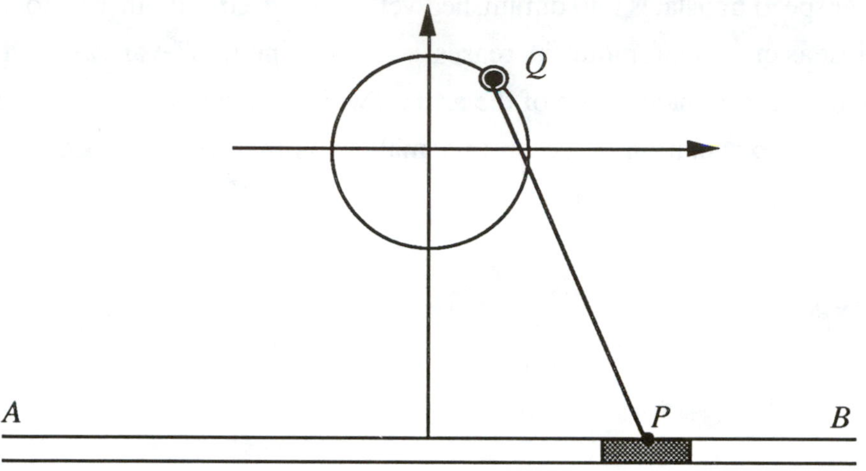

This project considers two questions related to the design of such an assembly. First, for a given shaft radius and connecting rod length, how far to the left and right would the valve cover move? Second, it may be important to estimate the stress on certain parts. In the case of the connecting rod, for example, one preliminary step would be to estimate the acceleration of the valve cover as it moves between A and B. This stress-acceleration connection seems plausible since force is proportional to acceleration and one of the main forces exerted on the connecting rod comes from its linkage to the valve cover.

Suppose the shaft rotates at the constant rate of 60 revolutions per minute, the radius of the shaft is 1 inch, the connecting rod is 4 inches long and the horizontal axis on which the valve cover moves is 3 inches below the center of the shaft. In particular, assume that the connecting rod length and the location of the valve cover axis are well matched in the sense that there is no binding as P passes back and forth between the third and fourth quadrants. Estimate the extreme left and right positions of P, the maximum absolute value of the valve cover's velocity and acceleration and the time at which each occurs. Also estimate the valve cover's velocity as P passes through the y-axis. Discuss any interesting or surprising features you observe in P's motion. How would it differ if the shaft's rotation were clockwise instead of counterclockwise? Give all answers in units involving inches and seconds accurate to the nearest hundredth at least.

Information for the instructor only:

Problem abstract: The goals of this project are the involvement of students in geometric and trigonometric modeling requiring use of a spliced function, the use of the computer and numerical estimation in performing max/min calculations that would be difficult to carry out analytically, the real-world interpretation of modeling results, the use of computer/calculator-generated graphs in developing a model and investigating its properties, the use of one-sided limits in determining limiting velocity and acceleration at a cusp, and the clear and carefully reasoned written or oral explanation of project work. While the problem statement does not pretend to be representative of the complexity involved in actual design, it is hoped that the element of realism will stimulate a sense of purpose and provide a context for motivating and reflecting on the meaning of a modeling effort.

Prerequisite skills and knowledge: This project can be assigned any time after students have been introduced to interpretation of first and second derivatives as velocity and acceleration and to the use of a software package capable of generating 2-dimensional graphs and enabling one to estimate with reasonable accuracy the maximum and minimum values of the first and second derivatives of a given function. However, without some additional knowledge (for example, familiarity with parametric equations), most students will find this project very challenging. It is helpful, but not necessary, to obtain a derivative formula (either by hand or through use of a computer algebra system) requiring the differentiation of sine and cosine functions and use of the chain rule.

Essential/useful library resources: none

Essential/useful computational resources: It is essential to have a software package that can be used to generate 2-dimensional graphs and to estimate with reasonable accuracy the maximum and minimum values of the first and second derivatives of a given function. It may be useful (but is not essential) to have software capable of obtaining simplified formulas for derivatives.

Example of an acceptable approach: It would be surprising if students did not experience some difficulty in getting the modeling underway. A likely (possibly with some instructor assistance) and constructive first draft effort at modeling the valve cover's motion would begin with the observation that t seconds after beginning from (1,0), the point Q in Figure 1 has coordinates $(\cos(2\pi t), \sin(2\pi t))$.

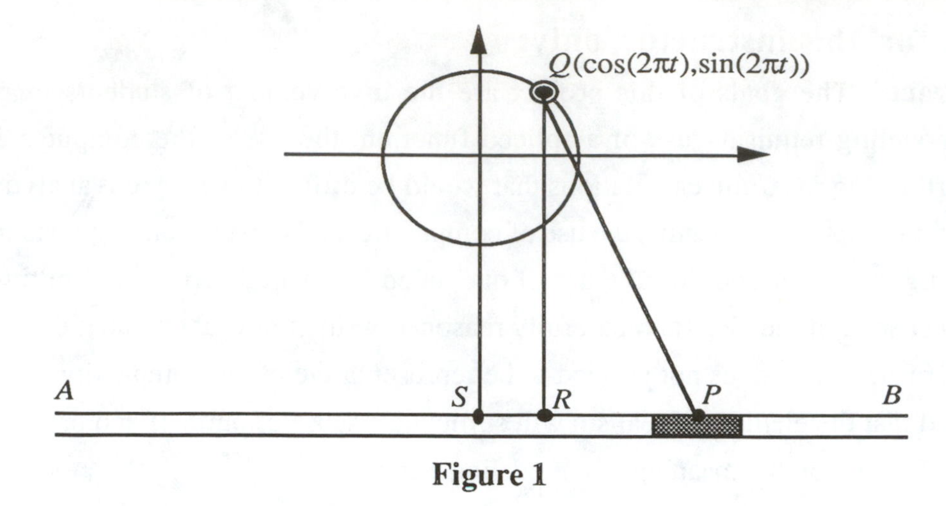

Figure 1

Since the figure given in the problem statement encourages focus on the fourth quadrant, it may at first seem that after t seconds the x coordinate of P in inches should be given by

$$x(t) = \text{length of } SR + \text{length of } RP.$$

Thus, using the Pythagorean Theorem, we have

$$x(t) = \cos(2\pi t) + \sqrt{16 - (\sin(2\pi t) + 3)^2}\ .$$

If this is the expectation, then the graph of $x(t)$ shown in Figure 2 should produce a surprise: P appears to stay entirely in the fourth quadrant moving back and forth between S and A.

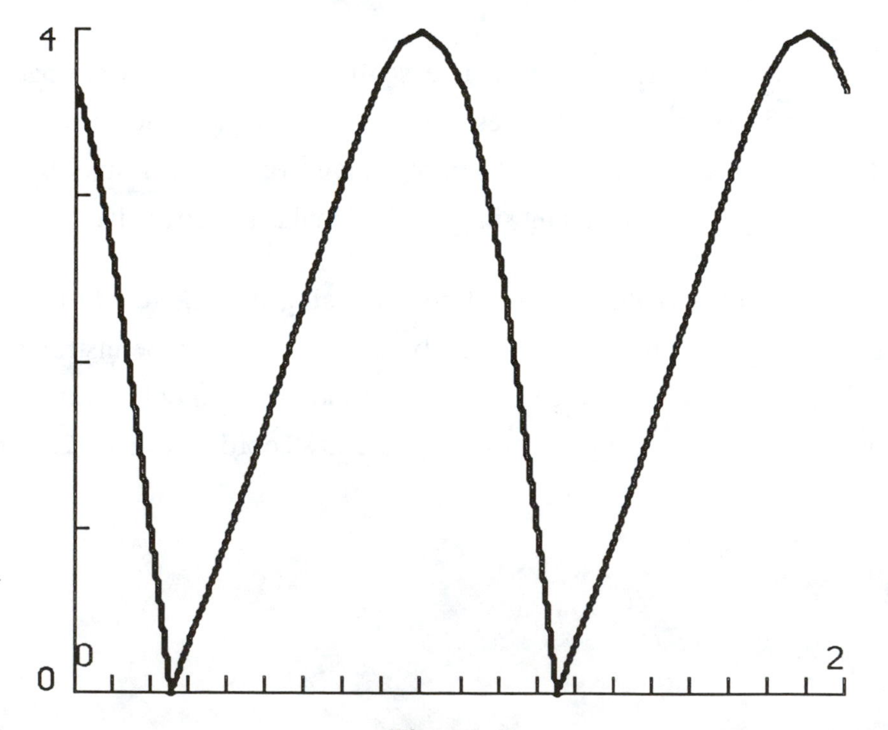

Figure 2

Reflection at this point should suggest to students that a different formula is required to describe the valve cover's motion when it passes into the third quadrant - the square root term should be subtracted rather than added. This gives rise to the following splice function; the domain shown is for the first full period given that Q begins at (1,0) with P in the fourth quadrant:

$$x(t) = \begin{cases} \cos(2\pi t) + \sqrt{16 - \sin(2\pi t) + 3)^2} & \text{for } 0 \le t \le 0.25 \text{ or } 1.25 \le t \le 2 \\ \cos(2\pi t) - \sqrt{16 - \sin(2\pi t) + 3)^2} & \text{for } 0.25 \le t \le 1.25. \end{cases}$$

A graph of the negative branch of this function (when the square root term is subtracted and P is held in the third quadrant) is shown is shown in Figure 3 and the superposition of the two, yielding a graph of the splice function, is given in Figure 4.

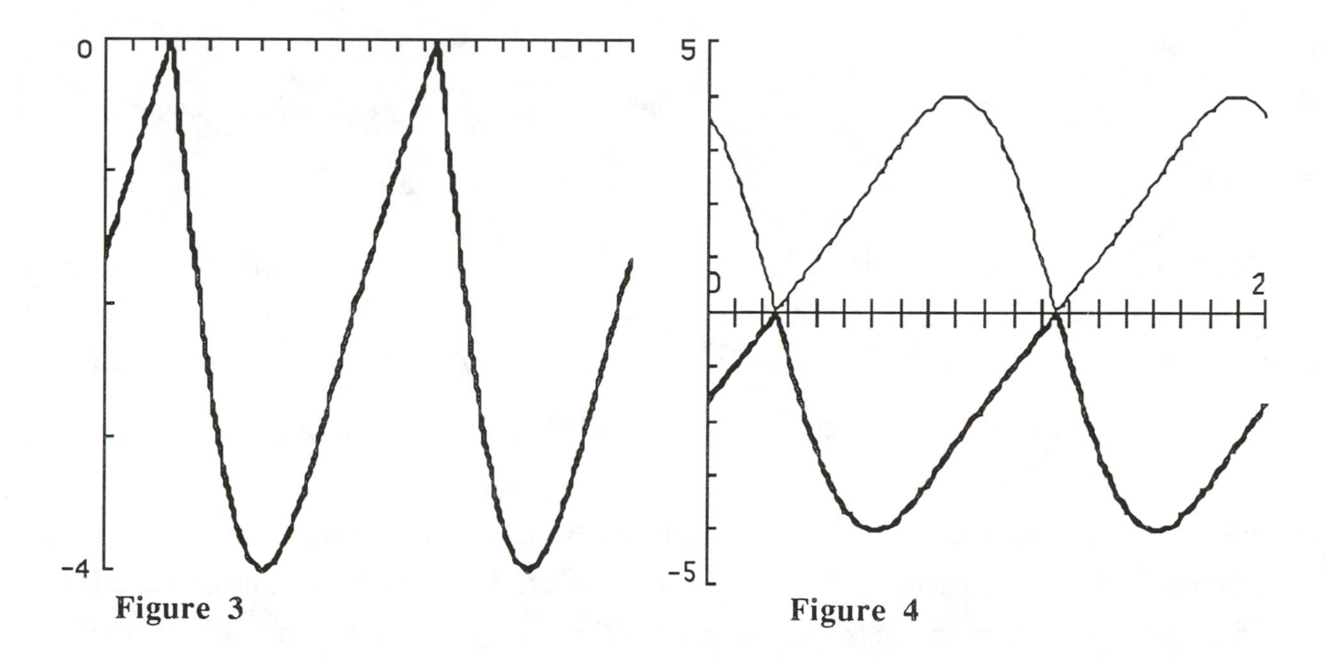

Figure 3 **Figure 4**

While finding extreme values for P's position, velocity and acceleration is straightforward in concept, carrying the calculations through by hand would be quite difficult. It is a good time to turn to an available software package for assistance. Extreme values of the position function (to the nearest ten thousandth) of 4 and –4 occur at t ≈ 0.8975 and 0.6025 respectively. Computer-generated graphs of velocity and acceleration for the positive branch (when the square root is added) and estimates of their extreme values are given below in Figures 5 and 6.

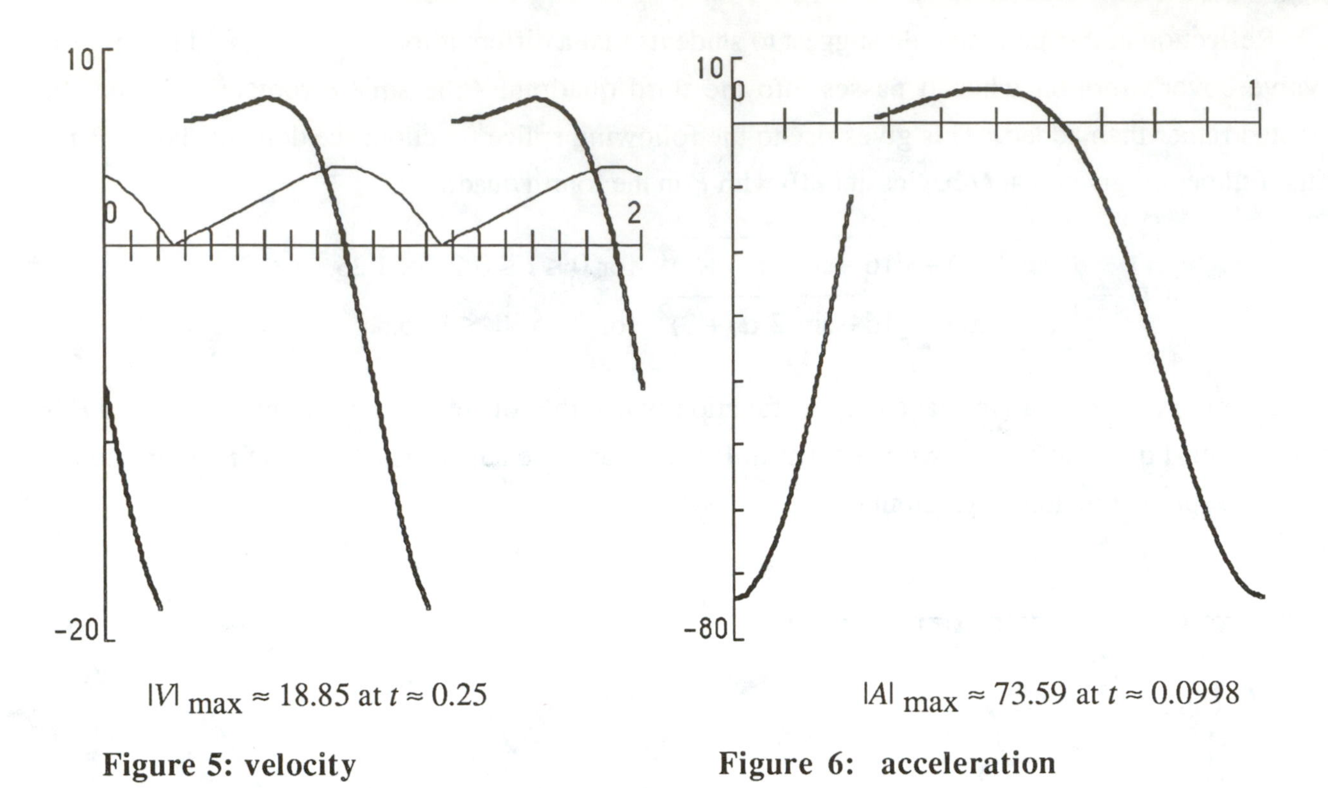

$|V|_{\max} \approx 18.85$ at $t \approx 0.25$

Figure 5: velocity

$|A|_{\max} \approx 73.59$ at $t \approx 0.0998$

Figure 6: acceleration

The velocity function in this model does not actually take on a maximum absolute value since, as Figures 2 and 5 suggest, the derivative of the positive branch is undefined at $t = 0.25$. However, with computer assistance, students should be able to find good approximations for the two one-sided limits:

$$\lim_{t \to .25^-} x'(t) \approx -18.8496 \approx -6\pi \text{ and } \lim_{t \to .25^+} x'(t) \approx 6.2832 = 2\pi.$$

Obtaining these figures by hand would be, as one would expect, relatively difficult. For the negative branch of the splice function, the situation with respect to extreme values is identical except that the absolute value of the acceleration is taken on at $t \approx .502$ and the values of the one-sided limits of the derivative at .25 are reversed.

Of special interest are the transition points between the two branches of the splice function; i.e., the points in time, $t = .25$ and $t = 1.25$, when P is passing through the y-axis. Figure 4 gives the appearance of a "smooth" transition from one branch to the other. If we let v_p and v_n denote velocity on the positive and negative branches respectively, then evidence for this "smoothness" can be found by discovering, with computer assistance, that

$$\lim_{t \to .25^-} v_p(t) = \lim_{t \to .25^+} v_n(t) \approx -6\pi \text{ and } \lim_{t \to 1.25^-} v_n(t) = \lim_{t \to 1.25^+} v_p(t) = 2\pi.$$

It may seem surprising that P's speed in passing through the y-axis is different when moving to the left as opposed to the right. A reason for this lack of symmetry can be discovered by visualizing P's movement through one entire period; in other words, through two revolutions of the shaft. In particular, note the difference in the orientation of the connecting rod when t approaches .25 from the left as opposed to when t approaches 1.25 from the left. This observation may prompt the conjecture that if the shaft were rotating clockwise rather than counterclockwise, then we should be able to observe that

$$\lim_{t\to .75^-} v_p(t) = \lim_{t\to .75^+} v_n(t) \approx -2\pi \text{ and } \lim_{t\to 1.75^-} v_n(t) = \lim_{t\to 1.75^+} v_p(t) = 6\pi.$$

That this is indeed the case may be discovered by computer- assisted investigation of the model in case the rotation is clockwise. Since clockwise rotation results in the point Q having coordinates $(\cos(-2\pi t), \sin(-2\pi t)) = (\cos(2\pi t), -\sin(2\pi t))$ after t seconds, the only alteration required in the splice function is to replace $\sin(2\pi t)$ wherever it occurs by $-\sin(2\pi t)$.

Title: The Tape Deck Problem

Author: Matt Richey, St. Olaf College

Problem Statement: Most cassette tape decks (and VCRs) have a counter which changes as the deck operates. However, most people are unaware of the relationship between the value of the counter and the time that the tape has been playing (the elapsed time). The purpose of this problem is to investigate this relationship.

Part I

Collect data and derive a formula expressing time elapsed (t) as a function of counter reading (x). That is, determine a function f so that $t = f(x)$ provides a reasonable "fit" to the collected data. You will need to give considerable attention as to how to measure a good "fit."

Part II

One can model the data more accurately if the derivative of time with respect to counter reading is considered. Recall that the derivative $\frac{dt}{dx} = f'(x)$ can be approximated by the change in t divided by the change in x: $f'(x) \approx \frac{\Delta t}{\Delta x}$.

Go back to your VCR to collect data and derive a formula expressing $\frac{\Delta t}{\Delta x}$ as a function of x. Now determine f and compare it to the function found in Part I. Which "fits" the data more accurately? Again, be certain to carefully describe how you have measured the accuracy of the "fits."

Information for the instructor only:

Problem abstract: The goals of this project are to investigate and develop a mathematical model of an everyday phenomenon. Ideas encountered can include empirical investigation and data collection, curve fitting, solving simultaneous equations, approximating the derivative, and solving and setting up differential equations. This problem is well-suited for empirical investigation. Since most students will either own or have easy access to a cassette deck or a VCR, many will probably recognize the problem. Most will have the initial impression that the relationship between the elapsed time and the counter is (more or less) linear. As it turns out, it is quadratic. This problem has the benefit that it presents a nonlinear phenomenon which is easily studied by first-year calculus students. The students can conclude empirically that the relationship is quadratic and then apply basic modeling techniques to explain why. The advantage of considering the derivative approximation in Part II is that the relationship at this level is in fact linear, so the curve fitting process is greatly simplified.

Prerequisite skills and knowledge: The prerequisites for this project are minimal. It can be done on an empirical level by students early in a calculus course. It also makes an excellent modeling project for students recently exposed to the derivative.

Essential/useful library resources: none

Essential/useful computational resources: A computer algebra system (CAS) would be quite useful but not necessary.

Example of an acceptable approach:

Part I:

The students should begin this problem by simply going out and collecting data. Looking at the time versus counter relationship, the following table of values should be representative of what they might obtain.

time = t	counter = x
0	0000
5	0374
10	0692
15	0973
20	1227
25	1462
30	1680
35	1886
40	2080
45	2265
end	2343

It is important at this point for the students to clearly state for what it is they are looking. The statement of the problem asks for the time, t, as a function of the counter reading, x. Thus, this data with t = time as a function of x = counter reading, and not the inverse relation, needs to be plotted accurately. Note that in this table, data was collected at a constant time interval. It would certainly be possible, and in fact better, to collect data at constant counter reading intervals. A plot of the data reveals that the relationship between the counter and the time is not linear. At this point, the students can begin to formulate some conjectures as to what form of curve best fits this data. A reasonable guess would be some sort of polynomial fit, although certainly other types of guesses are reasonable.

If a CAS is not available, then there are several ways to proceed. Since the relationship clearly is not linear, the next simplest place to look is at a quadratic fit. This means that the function takes the form $t(x) = Ax^2 + Bx + C$. The problem for the students will be to determine values for A, B and C. It will be interesting to see how students approach this problem. Most should notice that since $t(0) = 0$, it is fairly evident that an acceptable value for C is 0. The decision as to how to solve for the other values should be rather difficult for most students. One way to find appropriate values for A and B would be to use two pairs of points (x_1, t_1) and (x_2, t_2) and solve the simultaneous equations

$$t_1 = Ax_1^2 + Bx_1$$

$$t_2 = Ax_2^2 + Bx_2.$$

Solving for A and B can be a bit sticky, but it is not an extremely difficult task and it does give the students practice with simultaneous equations. Using, for example, the points (374,5) and (2080,40), one gets $A = .00000344373$ and $B = .0120678$. It would probably be prudent to repeat this procedure several times with different pairs of points in order to get several options for A and B. Doing so could possibly lead them into the study of matrix equations and linear algebra.

If the students have access to a computer algebra system which has some kind of curve fitting capability, then this would be a good place to use it. For example, using *Mathematica* with the data above and trying a quadratic fit, one gets

$$t(x) = .00469058 + .0120537x + .00000345143x^2.$$

If one tries a cubic fit, then the quadratic part of the answer is essentially the same and the coefficient of the x^3 term is on the order of 10^{-14}. Similar results hold for higher order fits. Thus with a CAS, the curve fitting part of the problem is quite simple.

Whatever the approach, the students should now verify that their function does in fact "do the job." Perhaps the best way for them to do this is to go back to the deck from which they got their data and check that the function does give acceptable values. What is "acceptable" is up to them, although they should be expected to discuss the term in their report. Since we are dealing with timing cassette tapes, an error on the order of one second is probably fine. In any case, this is a point with which the students should grapple. Another way to verify the correctness of the quadratic fit would be to consider the derivative $t'(x)$ and try to approximate it. If the relationship is in fact quadratic, then the derivative, of course, should be linear. This approach is, in fact, what is expected in Part II.

Part II:

As given in the problem statement, approximate the derivative with

$$f'(x) \approx \frac{\Delta t}{\Delta x}.$$

In order to obtain values for this approximation, they will need to choose a value for Δx which is small enough to yield a reasonable approximation to the derivative, but large enough to allow a reasonable measurement of Δt. For the example, with the cassette deck used earlier, a value of $\Delta x = 10$ works well. The results of their measurements should look something like the data in the following table.

x	$\frac{\Delta t}{\Delta x}$ (seconds/counter unit)
1-10	.731
100-110	.766
200-210	.810
300-310	.856
400-410	.908
500-510	.953
1000-1010	1.151
1100-1110	1.209
1200-1210	1.250
1300-1310	1.283
1400-1410	1.325
1500-1510	1.369
2000-2010	1.565
2100-2110	1.603
2200-2210	1.644

From a plot of this data it is not hard to believe that the relationship between $f'(x)$ and x is linear. There will naturally be more error of measurement here because of the difficulty of accurately timing these relatively short intervals. To approximate this with a line, the students will likely just choose two ordered pairs and find the slope and y-intercept for the line through these. Again, it would probably be a good idea to try various combinations in order to see which line works "best." The students could also carefully plot the table values on graph paper and hand-fit the "best" line through them. Also, some students might have knowledge of linear regression or have a CAS which will find the best line in that sense. In any case, the students should be able to get an approximation for $f'(x)$ of the form $f'(x) \approx ax + b$ or $f(x) = \frac{a}{2}x^2 + bx$. For the data above and using linear regression, one gets

$$f'(x) = .0004176188x + .7355622$$

or

$$f(x) = .0002088094\, x^2 + .7355622x.$$

Converting the units from seconds to minutes gives $f(x) = .000003480157x^2 + .01225937x$ which agrees quite closely with the previous results.

No matter how the equation for $f(x)$ is found, at this point the students must address the accuracy of their approximation. This is a valuable exercise because it introduces the students to the ideas of model verification and error. Using the data above and the form $f(x) = .000003480157x^2 +. 01225937x$, the students could obtain something like:

x = counter	t = actual time	$f(x)$ = projected time	error (minutes)	percentage error
0000	0	0	0	0
0374	5	5.072	.028	0.6
0692	10	10.150	.150	1.5
0973	15	15.223	.223	1.5
1227	20	20.282	.282	1.4
1462	25	25.362	.362	1.5
1680	30	30.418	.418	1.4
1886	35	35.500	.500	1.4
2080	40	40.556	.556	1.4
2265	45	45.621	.621	1.4
2343	end	47.829		

Notice that for this example the percentage error is consistently 1.5 or less. It is surprising how well a quadratic polynomial models this phenomenon.

Conjectures we expect that some students will make: It is hoped that students will engage the problem of choosing an appropriate function for a curve fit and will make a variety of conjectures about the data in this process. Some students will have had experience with linear regression and may go to that immediately in determining the time as a function of counter.

Questions for further exploration: A very interesting but much more difficult problem is to show that the relationship is, in fact, quadratic. A project which included addressing the problem could be done by strong students after some exposure to differential equations. A solution is included here.

The students must look more closely at how a tape deck operates. A rough schematic of the essentials of a tape deck appears in Figure 1.

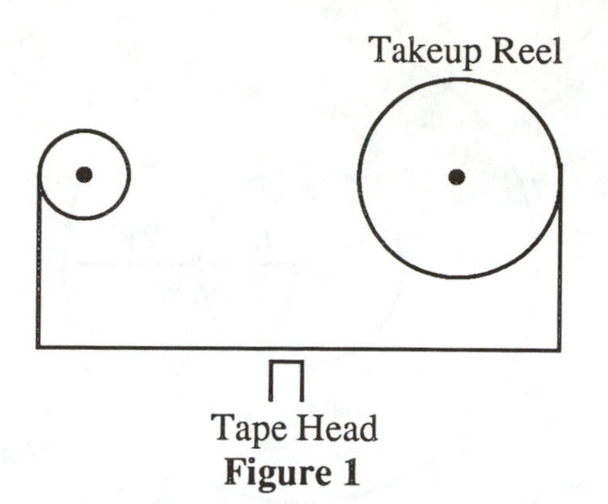

Figure 1

Research on the students' part should reveal that the key facet of the operation of a tape deck is that the tape must be pulled across the tape head at a constant rate of $1\frac{7}{8}$ inches per second. Thus the two reels must turn at varying rates during the course of a tape being played. Early on, when most of the tape is still on the reel on the left, the right wheel (the take-up reel) turns rather quickly since it has a relatively small radius. Later on, when more time has passed, the radius of the take-up wheel is larger and hence it can turn much more slowly. Thus the rate of rotation mirrors the rate of change of x as a function of t, i.e., when t is small, the rate is large and when t is large the rate is small. This might lead one to guess that the counter is related to the turning of the take-up reel. This turns out to be the case. Some observation should reveal that the counter effectively counts the revolutions of the wheel. More precisely, if w represents the number of revolutions of the wheel, then $x = kw$ where k is a constant dependent on the deck. This relationship can be discovered very simply, for example, by more data collection and a plot of x versus w. Therefore the question can be rephrased as: Why is the time t quadratic in w?

It is also necessary to incorporate into the problem the constant tape speed of $1\frac{7}{8}$ inches per second. Since the tape is moving at a constant rate, the length of tape on the reel is proportional to the time elapsed. Thus the question could be rephrased again as: Why is the length L of tape on the take-up reel quadratic in the number of revolutions w? This rephrasing of the question is essential. In all likelihood, most students will not discover it on their own. If so desired, the instructor could lead them through this argument either in a class discussion or with a handout.

There are at least two ways to show why the length L is quadratic in w. One way is to observe that the length of the tape times the thickness of the tape is simply the area A of the tape on the reel. Letting R_0 be the radius of the hub of the take-up reel and R_1 be the width of the section of the reel with tape on it, this area is $A = \pi((R_1 + R_o)^2 - R_o^2) = \pi(R_1^2 + 2R_1R_o)$ (Figure 2).

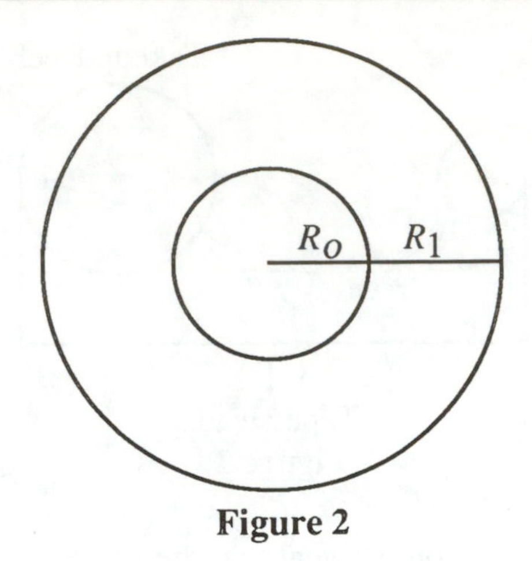

Figure 2

Now R_1 is just the thickness of the tape τ times the number of revolutions w. Putting all of this together, one gets

$$L = A/\tau = (\pi(w\tau)^2 + 2(w\tau)R_o)/\tau = \pi\tau w^2 + 2\pi R_o w.$$

Another way to derive the quadratic relationship is to model the derivative $\frac{dL}{dw}$. This approach could serve as an introduction to differential equations. It would fit in nicely if the students have already approximated the derivative as a way of finding a formula for $f(x)$. If they have done so, then it would be natural to investigate why $\frac{dt}{dx}$ is linear in x or equivalently why $\frac{dL}{dw}$ is linear in w.

One way to begin is to ask how a little change in rotation affects the change in L, that is, find ΔL versus Δw. One might notice that as the wheel turns, more tape is added on to the outer edge, i.e., the circumference. We obtain the approximation

$$\Delta L \approx 2\pi R(w)\Delta w$$

where $R(w)$ is the radius of the entire reel after w revolutions (Figure 3).

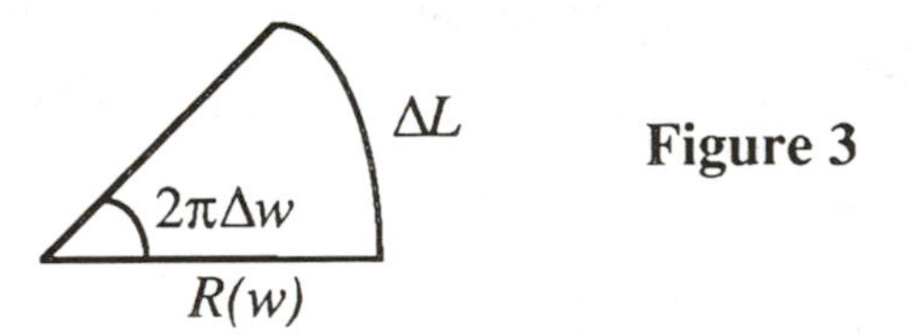

Figure 3

This is a reasonable approximation because if Δw is small, then $R = R(w)$ is essentially constant. In the above notation $R = R_1 + R_o$ and so $R = \tau w + R_o$. Thus

$$\frac{\Delta L}{\Delta w} \approx 2\pi(\tau w + R_o)$$

or, letting $\Delta w \to 0$,

$$\frac{dL}{dw} = 2\pi(\tau w + R_o).$$

From this one gets $L = \pi\tau w^2 + 2\pi R_o w$ (note that the constant of integration must be 0). Since $t = L/S$ where $S = 1\frac{7}{8}$ inches per second and $w = x/k$ where k is a constant of proportionality, we obtain

$$t(x) = \frac{\pi\tau}{Sk^2}x^2 + \frac{2\pi R_o}{Sk}x$$

as the equation relating t and x. In both derivations of $t(x)$ there are several assumptions. For example, one must assume that the speed of the tape is *exactly* $1\frac{7}{8}$ inches per second. Surely it isn't, all tape decks vary the speed somewhat (the wow and flutter). Also, it is assumed that the tape winds up on the reel with constant compression. That is, the tape near the center is not wound any more tightly than the tape near the outer edge.

Special implementation suggestions: Beware: some of the newest tape decks and VCR models use an electronic counter which is related to time elapsed in a linear fashion. Students should be led away from such machines.

ANTIDERIVATIVES AND DEFINITE INTEGRALS (PRE-FUNDAMENTAL THEOREM)

The projects in this section have been separated from the following section primarily because of the break that occurs between semesters or quarters in most calculus sequences. The first two projects of this section, *Population Growth* and *Drug Dosage,* are the only ones to use antiderivatives directly, although the last project, *A Fundamental Project*, gives the students the opportunity to discover the relationship between antiderivatives and area by means of "area functions." In both *Population Growth* and *Drug Dosage*, students need to be able to solve $y' = ky$. Thus, appropriate placement in a specific course will depend on when students in that course are familiar with the derivative of the natural exponential function. The other three projects, *Logarithms: You Figure It Out, Numerical Integration and Error Estimation,* and *An Integral Existence Theorem* all involve students deeply in Riemann sums and/or numerical integration. These projects are appropriate for late in a Calculus I course or early in Calculus II.

Title: Population Growth

Author: Wayne Roberts, Macalester College

Problem Statement: It is common to assume that a population will grow at a rate that is proportional to its size; that is, the larger the population, the larger its growth rate. If we denote population size by p and time by t we have a population growth model $\frac{dp}{dt} = kp$ (model 1). Solutions to this differential equation are of the form $p(t) = p_o e^{kt}$, where p_o is the initial ($t = 0$) population. The constant k is assumed to be a characteristic of the population under consideration, higher for rabbits than for people, for example. Used in the context of human population growth, model 1 is called the Malthusian Law, in honor of Thomas Malthus (1766-1834), author of an influential essay on overpopulation.

A second model of population growth was published by Raymond Pearl and Lowell Reed. They contended that it was absurd to try to predict population with any equation whose value continues to increase without bound; that the population of any confined geographic area must have an upper bound, say M, beyond which the population will not grow. Specifically, they hypothesized certain conditions for a satisfactory model:

1. Asymptotic to a line $p(t) = M$ when $t \to +\infty$.
2. Asymptotic to a line $p(t) = 0$ when $t \to -\infty$.
3. A point of inflection at some point $t = \alpha$ and $p = \beta$.
4. Concave upward to the left of $t = \alpha$ and concave downward to the right of $t = \alpha$.
5. Slope is never zero for any value of t.
6. Values of $p(t)$ varying continuously from 0 to M as t varies from $-\infty$ to $+\infty$.

A mathematical model which fits these conditions can be derived along the lines of model 1. Assume that there is an upper bound M beyond which the population will never grow, and that the rate of growth of $p(t)$ is proportional to the product of the population p and the difference $\frac{M-p}{M}$. (Note that this term will have very little influence for small values of p as it will be near 1 but will have a dampening effect for p near M as it will be near 0.) Model 2 can thus be written as $\frac{dp}{dt} = kp\left[\frac{M-p}{M}\right]$. Note that when the population is small, the population growth rate is close to that of model 1, but as the population size gets close to M, its growth rate becomes very small. Solutions to this differential equation are of the form $p(t) = \frac{Mp_o}{p_o + (M - p_o)e^{-kt}}$. Pearl and Reed used the equivalent equation $p(t) = \frac{be^{at}}{1 + ce^{at}}$ in their model.

Below is a set of questions concerning each model. Your task is to answer these questions and submit a written report of your findings.

(A) Model 1:

(1) Verify that the equation for $p(t)$ in model 1 does satisfy the differential equation for model 1.

(2) Use the data in Table 1 below to find a good choice for the constant k in model 1. Be careful to explain how you used the data in the table, what assumptions you made, and describe any limitations of your equation for $p(t)$ under model 1.

(3) Make a new column in table 1 listing predicted population under model 1. Describe how well your version of model 1 fits the population data from table 1.

(4) According to model 1, what will be the population of this country in 2000? in 2010?

(5) How well does model 1 work and why?

(B) Model 2:

(6) Verify that the equation for $p(t)$ does satisfy the differential equation.

(7) Verify that the two forms of the equation for $p(t)$ are equivalent.

(8) Pearl and Reed, beginning at 1780 and using the data for 1790, 1850 and a different figure for 1910 (91,972,266), published a population equation, $p(t) = \dfrac{2930300.9}{0.014854 + e^{-0.0313395t}}$. See if you can obtain the same equation using the same data points. Then check to see how well this function fits the data.

(9) Pearl and Reed came to the conclusion that the maximum population for our country is 197,274,000. How do you think they got this value? Is it correct?

(10) Pearl and Reed identified 1914 as the year in which the U.S. population curve would turn from concave up to concave down. How do you think they reached this conclusion?

(11) Using the data that has accumulated since Pearl and Reed did their work, try your hand at finding the parameter values to best fit model 2 to the data. State your resulting population equation, your predicted maximum population and the year in which the concavity of the population curve changes.

Table 1: U.S. Population (taken from *The World Almanac*, 1992)

year	population	year	population
1790	3,929,214	1890	62,979,766
1800	5,308,483	1900	76,212,168
1810	7,239,881	1910	92,228,496
1820	9,638,453	1920	106,021,537
1830	12,860,702	1930	123,202,624
1840	17,063,353	1940	132,164,569
1850	23,191,876	1950	151,325,798
1860	31,443,321	1960	179,323,175
1870	38,558,371	1970	203,302,031
1880	50,189,209	1980	226,542,203
		1990	248,709,873

Information for the instructor only:

Problem abstract: This project gives students an opportunity to consider a realistic application of mathematics to an important social issue. The notion of fitting a curve to a set of data and using the resulting equation of the curve to predict other (future) values of the data will be a new one for most students.

Prerequisite skills and knowledge: This project can be assigned any time after students have been introduced to antiderivatives and the solution of separable differential equations.

Essential/useful library resources:

(1) "On the rate of growth of the population of the United States since 1790 and its mathematical representation," *Proceedings of the National Academy of Sciences*, Vol. 6, No. 6, June 15, 1920, pages 275-288.

(2) A section on population growth in any calculus text.

Essential/useful computational resources: A CAS with computational and graphing capabilities would be helpful to the students.

Example of an acceptable approach: I address below only those questions that are not straightforward.

(2) If we choose 1790 as the beginning ($t = 0$), and try 1900 ($t = 110$), we get $3.9 = p_o e^o$, and $76.2 = p_o e^{110k}$. It follows that $p_o = 3.9$ and $e^k = 1.0274$; so $p(t) = 3.9\ (1.0274)^t$. Or if we choose 1850 ($t = 60$) and 1910 ($t = 120$), we get $23.2 = p_o e^{60k}$ and $92.2 = p_o e^{120k}$. If we note that $e^{60k} = \dfrac{23.2}{p_o}$, then we can say that $92.2 = p_o [e^{60k}]^2 = p_o \left[\dfrac{23.2}{p_o}\right]^2$. It follows that $p_o = 5.84$. We then find that $e^k = 1.023$, so $p(t) = 5.84\ (1.023)^t$. Obviously, students may make other choices, but you should be sure that they consider what might be better choices of data points.

(3) Here, it would also be useful if they plot their curve and the data points on the same coordinate system for comparison. Their new table should have the following headings:

<u>year</u> <u>population</u> <u>model 1</u>

(4) Using the second equation in (2), we get for the year 2000 $p(210) = 692.3$ million people. For 2010, we get $p(220) = 869.1$ million people.

(5) Model 1 doesn't work very well because it grows very rapidly (hence the term "exponential growth" in common language) with an ever-increasing rate of growth .

(7) Multiply the numerator and denominator of the first equation by e^{kt}.

Then $a = k$, $b = \dfrac{Mp_o}{M - p_o}$, and $c = \dfrac{p_o}{M - p_o}$.

(8) When students try to solve the resulting system of three equations in three variables , they will find it very challenging, even with the use of most CAS's. They should at least check the given data

points to see if the given parameter values do give population results very close to the tabulated values ...they do. To solve the resulting system of equations:

$$3{,}929{,}214 = \frac{b}{c + e^{-10a}},$$

$$23{,}191{,}876 = \frac{b}{c + e^{-70a}},$$

$$91{,}972{,}266 = \frac{b}{c + e^{-130a}},$$

it is best to first choose two pairs and for each pair, divide one by the other, thus forming a system of two equations in variables a and c, eliminating b. Then after some algebraic manipulation, divide one of those by the other, eliminating c and yielding one equation in a, which can be solved using logarithms after some algebraic manipulation. The rest is relatively straightforward algebra. The result of my calculation was $p(t) = \dfrac{2930499}{0.0148533 + e^{-0.0313383t}}$, which is reasonably close to Pearl and Reed's result and almost fits the three data points exactly. The system can be solved with Maple V yielding $p(t) = \dfrac{2930499.0}{0.01485326825 + e^{-0.03133829349t}}$. This fits the data exactly.

(9) They divided the numerator and denominator of $p(t)$ by 0.014854 and then evaluated the limit of $p(t)$ as t increases without bound.

(10) I'm not sure how they found it, but I would encourage students to calculate the second derivative and find its roots. A CAS may be helpful in calculating the derivatives and finding any roots of the second derivative, but Maple, at least, needed some help. Clearing the denominator of the second derivative equation first does the trick. Maple then yields 134.3 (1914!). Alternatively, use of the "evalf" function on the second derivative at specific times near 1914 ($t = 134$), also supports Pearl and Reed's claim. It is quite useful to plot the function to scale and see where it seems to change concavity...this may be what Pearl and Reed did, by hand. It plots quite nicely with Maple. Do you have any better ideas?

(11) Again, the results will depend on the data points students choose to find values for the parameters in model 2. They should show a plot of their version of model 2 with the actual data points from the table. To answer the other parts of this question, students should use some of the techniques developed in answers to previous questions.

Conjectures we expect that some students will make: We hope that most would recognize the large number of factors that could be considered in this sort of modeling.

Questions for further exploration:

(12) Suppose that your projected maximum population were reached. What would be the population

density of our country and how would it compare to the population density of other countries?

(13) Pearl and Reed included arguments about the number of calories needed to sustain various populations, the food calories produced in various countries, etc. How have their arguments held up? Can they be modified to look reasonable again?

Title: Drug Dosage

(The idea for this project came from UMAP module unit 72, "Prescribing Safe and Effective Dosage" by Brindell Horelick and Sinan Koont.)

Authors: Diane Schwartz, John Maceli, Eric Robinson, Stan Seltzer, and Steve Hilbert, Ithaca College

Problem Statement: The concentration in the blood resulting from a single dose of a drug normally decreases with time as the drug is eliminated from the body. In order to determine the exact pattern that the decrease follows, experiments are performed in which drug concentrations in the blood are measured at various times after the drug is administered. The data are then checked against a hypothesized function relating drug concentration to time.

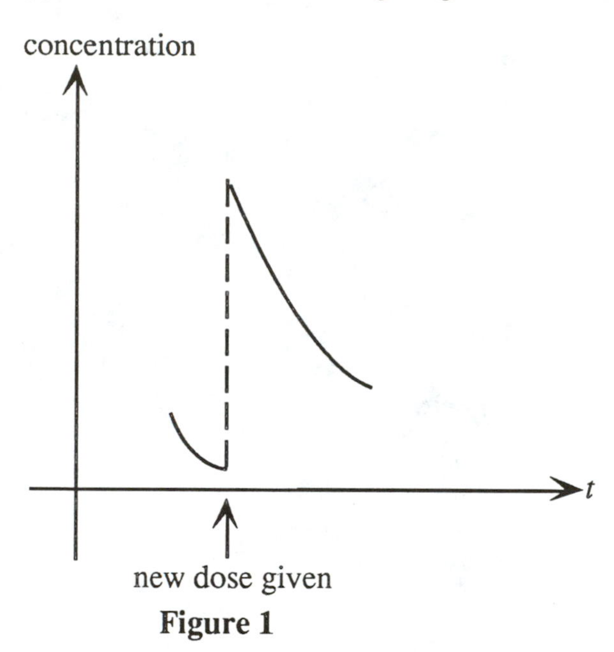

Figure 1

[At this point we make a simplifying assumption: namely that when the drug is administered, it is diffused so rapidly throughout the bloodstream that, for all practical purposes, it reaches its fullest concentration instantaneously. Thus, the concentration jumps from a low or zero level to a higher level in zero time. Visually, this represents a vertical jump on the graph of concentration vs. time. (See Figure 1) This assumption is probably nearly justified for drugs that are administered intravenously, for example, but not for drugs that are taken orally.]

The simplest function to hypothesize as a model of drug concentration is a linear one: that is, we might start by assuming that the concentration of the drug in the blood is a linear function of the time since the dose was administered.

Suppose a single dose of a certain drug is administered to a patient at time $t = 0$, and that the blood concentration is measured immediately thereafter, and again after four hours. The results of two such experiments are given in Table 1.

Data:	**Experiment 1**	**Experiment 2**
Concentration at time $t = 0$:	1.0 mg/ml	1.5 mg/ml
Concentration after 4 hours:	0.15 mg/ml	0.75 mg/ml

Table 1

A. For this part, assume that the function describing concentration as a function of time is linear. Each data set in Table 1 represents a different drug and a different initial dose. For *each* data set:

1) Sketch a graph of the concentration function, that is, graph the level of concentration vs. time. Assume concentrations are measured in milligrams per milliliter, and time is measured in hours.

2) Predict the time when the blood becomes free of the drug, assuming no further doses are administered.

3) Describe the rate at which the drug is eliminated. Does the rate of elimination seem to depend on any other quantity (e.g. level of concentration)?

4) Predict what the graph of concentration level vs. time would look like if further doses of the drug were administered every six hours for forty-eight hours.

5) Predict what would happen to the concentration level of the drug if it were administered every six hours indefinitely.

B. Now assume that the rate at which the concentration is decreasing at time t is proportional to the concentration level at time t. This idea can be modeled by a *differential equation*, namely:

$$\frac{dy}{dt} = -ky$$

where y is the concentration of the drug in the blood at time t, and k is a constant. Using the same data sets as in part A, solve the differential equation, and answer questions 1 to 5 above. [In fact, this model, Model B, has been shown in clinical tests to be the more accurate one.]

C. A problem facing physicians is the fact that, for most drugs, there is a concentration below which the drug will be ineffective and a concentration above which the drug will be dangerous. [See Figure 2.]

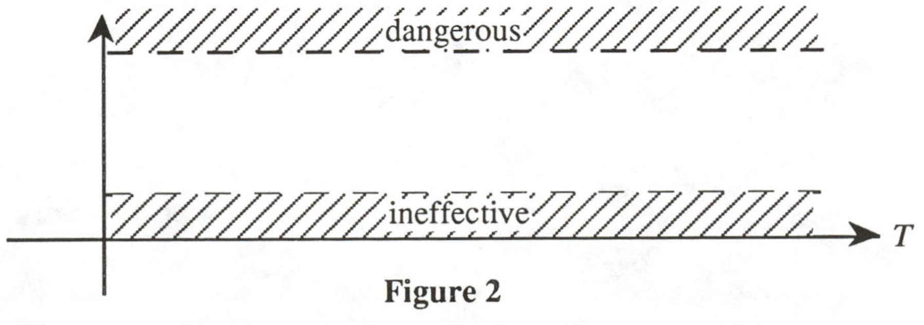

dangerous

ineffective

T

Figure 2

1) Suppose that for the drug in experiment 2, the minimum effective level is 0.45 mg per ml and the maximum safe level is 2.15 mg per ml. If the dose in the experiment is given every six hours, will the appropriate concentrations be maintained? Indefinitely? Explain. If the answer is no, can you achieve a satisfactory long-run level just by adjusting the time between doses? Just by adjusting the dose? (Assume there is a simple way to tell just how much substance must be administered in order to raise the concentration by any given amount. That is, you can answer this question by specifying how much the concentration of drug in the blood needs to be raised by each dose.)

2) Answer the same questions assuming the minimum effective level is 0.5 mg per ml, and the maximum safe level is 1.65 mg per ml.

Information for the instructor only:

Problem abstract: This project deals with functions as models of drug concentration levels. The students deal with the functions graphically and in terms of upper and lower bounds, not as formulas to be mechanically manipulated. Also, this project introduces numerical series in a concrete context. Convergence and divergence are introduced by constructing graphs of dosage levels, whose values at the times when dosages are given are the partial sums of geometric series. The issues of convergence and the limit of a converging series become meaningful in the context of avoiding dangerous drug levels.

Prerequisite skills and knowledge: Familiarity with the exponential function. Ability to solve $y = -ky$. Geometric series. The students will be able to *compute* the limit of the geometric series if they have that skill. The alternative is to have them observe the apparent limit by looking at the graph of the sequence of partial sums (which they are required to construct anyway).

Essential/useful library resources: none

Essential/useful computational resources: none

Example of an acceptable approach:

A. Linear model.

1) The students should draw a segment of a straight line connecting the given two points. The graph shouldn't extend below the t-axis.

2) For each experiment, the students could simply set the linear function equal to 0 and solve. For experiment 1, they should obtain $t = 4 / (0.85) \approx 4.7$. For experiment 2, $t = 8$. These values could also be predicted from a carefully drawn graph.

3) The rate at which the drug is eliminated is simply the slope of the line segment. This rate is constant (hence independent of other quantities) until the drug is eliminated. This question may cause some students to question this linear model.

4) In experiment 1, the drug is completely eliminated before the next dose is administered. Hence, the students' graphs should have period 6.

In experiment 2, the drug is not eliminated before the next dose is administered, so each line segment starts 0.375 units higher than the previous one. Hence, the graph of the function satisfies $f(t + 6) = f(t) + 0.375$.

5) In experiment 1, the function is periodic with period 6.

In experiment 2, the concentration level would grow indefinitely.

B. For experiment 1, $y = \exp\left(\frac{t}{4}\ln(0.15)\right)$.

For experiment 2, $y = 1.5\exp\left(\frac{t}{4}\ln(0.5)\right)$.

1) By hand or by computer, the students should be able to sketch these easily.

2) Hopefully, some students will change the question to predict when the concentration level goes below some measurable threshold.

3) The students should all realize the rate is varying with the level of concentration. Hopefully, they will realize the answer is furnished by the given differential equation, $y' = -ky$.

4) and 5) For both experiments, since the drug is never eliminated during the 6-hour period between doses (in experiment 1, about 94.2% is eliminated; in experiment 2, about 47% is eliminated), each peak amount (amount immediately after a new dose is administered) is higher than the previous peaks. However, the concentrations at the peaks are the partial sums of a geometric series. For example, in experiment 2, the peaks converge to approximately 2.32.

If the students have not yet seen series, this problem should serve as excellent motivation for the concepts of partial sums and convergence.

C. 1) In the original experiment 2, with a new dose every 6 hours, the treatment is always effective (the lowest value as $t \to 6^-$ is approximately 0.53 mg/ml). However, since the peaks approach 2.32, this treatment is unsafe.

There are several approaches the students might take to adjusting the time between doses. One method would be to calculate when the initial dose decreases to the 0.5 mg/ml level and readminister then. Next, calculate when this new concentration of 2 mg/ml will decrease to 0.5 mg/ml. Thus, after the first cycle of treatment, one can make the concentration periodic with values in the range 0.5 to 2.0.

If the student tries for all the time intervals, including the first, to be the same length, the interval of $t = 4\ \ln(0.3) / \ln(0.5) \approx 6.948$ works. The lowest concentration (as $t \to 6^{-}$) is 0.45 and the peaks converge to approximately 2.14. If the students try this method, they will almost surely need to know how to compute the limit of a geometric series, since the limiting value of the peaks is so close to 2.15.

If the students try to just adjust the dose, they should be successful. For example, a dosage of about 1.3 mg/ml will have a minimum value above 0.45 and the peak values will converge to about 2.011.

2) Since the range of acceptable values has width of only $1.65 - 0.5 = 1.15$, it's impossible to have an acceptable dosage pattern by just adjusting the time (leaving the dosage at 1.5). It is possible to adjust the dosage if the first dose is larger than the others. For example, a first dose of 1.6 mg/ml, followed by booster doses of approximately 1.0343 mg/ml will keep the concentration level in the 0.56 to 1.6 range.

Questions for further exploration: The following question is interesting and quite open-ended. Arguments may be based on statistical techniques such as hypotheses testing.

D. A patient in a certain clinic has been taking the drug XL37 regularly for several months. He has experienced some health problems which he attributes to the drug. The maximum safe level of XL37 in the blood stream is known to be 2.0 mg per ml. The patient has been tested periodically for concentration of XL37 in his blood, and the level found has been consistently between 1.9 and 1.95 mg per ml. The patient wishes to initiate a malpractice suit against the clinic, claiming that he has been administered an overdose of the drug.

1) Taking the point of view of the lawyer defending the clinic, argue why the patient's claim is not correct.
2) Taking the point of view of the attorney for the patient, argue why there is evidence of an overdose.
3) As judge in the case, what further information would you like to see before deciding this case?

Special implementation suggestions: There is a lot of merit in giving this project *before* introducing series of any kind. This project serves as a good introduction to the amazing idea that an infinite series can have a finite sum.

Title: Logarithms: You Figure It Out

Author: Matt Richey, St. Olaf College

Problem Statement: One often sees the natural logarithm function defined as

$$\ln x = \int_1^x \frac{1}{t}\,dt.$$

Suppose that all you know about the definite integral is that it is defined as the limit of Riemann sums and represents the (signed) area under the curve. Your project is to derive as many of the standard properties of the logarithm function as you can using only its definition as an integral and properties that you derive.

Following is a sketch of the derivation of the formula $\ln a^{-1} = -\ln a$ using the above definition: We must show $\int_1^{a^{-1}} \frac{1}{t}\,dt = -\int_1^{a} \frac{1}{t}\,dt$. Rewriting yields $\int_1^{1/a} \frac{1}{t}\,dt = \int_a^{1} \frac{1}{t}\,dt$. For geometric clarity we assume $a < 1$ and consider the following picture:

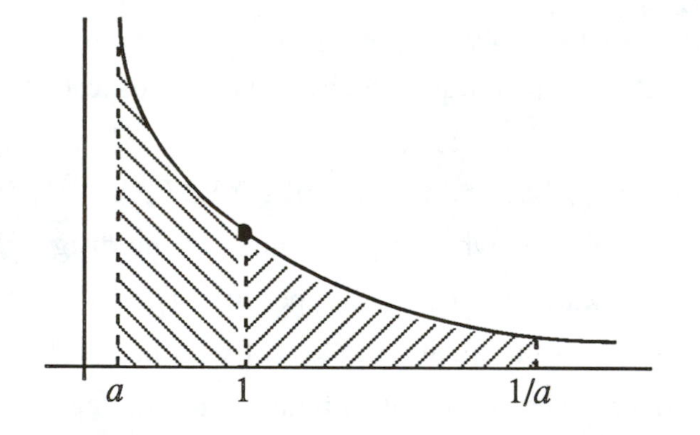

The left-hand side of the equation corresponds to the area on the right and the right-hand side to that on the left. We calculate a Riemann sum using n equal subintervals and left-hand endpoints for each interval.

$\int_1^{1/a} \frac{1}{t}\,dt$: Subinterval widths are $\dfrac{\frac{1}{a}-1}{n}$.

Function evaluation points are $1 + j \cdot \dfrac{\left(\frac{1}{a}-1\right)}{n}$ for $j = 0, \ldots, n-1$.

The corresponding Riemann sum is $R_1 = \displaystyle\sum_{j=0}^{n-1} \left(\frac{\frac{1}{a}-1}{n}\right) \cdot \frac{1}{1 + j \cdot \frac{(\frac{1}{a}-1)}{n}}$.

$\int_a^1 \frac{1}{t}\,dt$: Subinterval widths are $\frac{1-a}{n}$.

Function evaluation points are $a + j \cdot \frac{(1-a)}{n}$ for $j = 0,\ldots,n-1$.

The corresponding Riemann sum is $R_2 = \sum_{j=0}^{n-1} \left(\frac{1-a}{n}\right) \cdot \frac{1}{a + j \cdot \frac{(1-a)}{n}}$.

Part I: Complete the above proof by:

a) Verifying that the two Riemann sums R_1 and R_2 are in fact the same and explain the significance of this,

and

b) Discuss the case $a \geq 1$.

Part II: Derive as many of the standard properties of the logarithm function as you can using only the definition of a Riemann sum.

Suggestion: Before attempting to prove a general relationship, it might be useful to try out your method using specific values for a, b, and n.

Information for the instructor only:

Problem abstract: The goals of the project are to gain familiarity with Riemann sums and with the logarithm function. Also, this project will give students experience with developing and writing convincing mathematical arguments. This project gives students a chance to work with two apparently dissimilar and often troublesome concepts, logarithms and Riemann sums. There is ample opportunity for them to investigate the relationship between these topics using examples and special cases. Also, there are several ways to derive these properties from the definition above. Finally, the students are expected to present convincing arguments (proofs) for the validity of the familiar properties of the logarithm. This should give them experience writing about mathematics. This project could easily be done in two weeks or less in groups of 2 or 3 students.

Prerequisite skills and knowledge: An understanding of the elementary theory of integration up through Riemann sums. The Fundamental Theorem of Calculus is not needed. This project could be done in a first semester course right after the integral is introduced, or early in the second semester.

Essential/useful library resources: none

Essential/useful computational resources: none

Example of an acceptable approach: Part I is included simply to get the students on the right track. In a), one simply multiplies the terms in the sum R_1 by $\frac{a}{a}$ to yield the terms in the sum R_2. Students should then discuss that the two integrals must be equal if their upper sums are equal for all n. In b), it is probably sufficient to note the following picture indicates that the case $a > 1$ is analogous to $a < 1$.

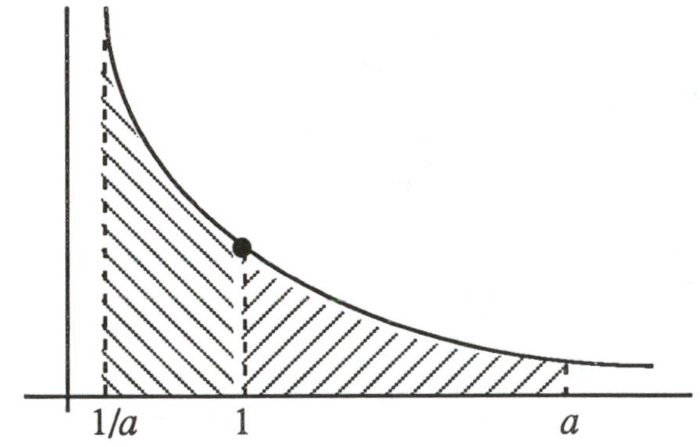

If one wants to be slick, just set $b = \frac{1}{a}$ and apply case 1 to b.

Part II is where the real work begins. The "standard" properties of the logarithm that the problem refers to include:

$$\ln ab = \ln a + \ln b$$

$$\ln a/b = \ln a - \ln b$$
$$\ln a^b = b \ln a$$

where $a, b > 0$. Most students will probably start with the first property, since it is one with which they are most familiar. Written in terms of integrals it becomes

$$\int_1^{ab} \frac{1}{t}\, dt = \int_1^{a} \frac{1}{t}\, dt + \int_1^{b} \frac{1}{t}\, dt.$$

The key observation is that for any $a, b > 0$

$$\int_1^{ab} \frac{1}{t}\, dt = \int_1^{a} \frac{1}{t}\, dt + \int_a^{ab} \frac{1}{t}\, dt.$$

Hence one must prove

$$\int_a^{ab} \frac{1}{t}\, dt = \int_1^{b} \frac{1}{t}\, dt.$$

Some students might have a difficult time seeing that this is the way to rephrase the question. However, if they are encouraged to pick values for a and b and to draw pictures, then it is not unreasonable to expect that they will discover this formulation.

If they have not done so already, now is a good time for students to choose some specific values for a and b. For example, with $a = 3$ and $b = 5$, the problem is to show that for $y = \frac{1}{x}$, the area under the curve from $x = 1$ to $x = 5$ is the same as the area under the curve from $x = 3$ to $x = 15$. These areas seem to be very close if not, in fact, equal (the second one is 3 times as long but only 1/3 times as tall). Since the students are told that area is defined using Riemann sums, they will probably calculate a Riemann sum for each region. A standard way to do so would be to use n equal subdivisions of each interval and to pick the left-hand endpoint to determine the height of each rectangle. Using the intervals [1, 5] and [3, 15] and 5 subintervals, it is easy to see that in the first case that $\Delta x = 1$ while in the second case $\Delta x' = 3$. For the interval [1, 5], one gets the partition $1 < 2 < 3 < 4 < 5$ and a Riemann sum of $1 + \frac{1}{2} + \frac{1}{3} + \frac{1}{4} + \frac{1}{5}$. For the interval [3, 15], one gets the partition $3 < 6 < 9 < 12 < 15$ and a Riemann sum of $(3)\frac{1}{3} + (3)\frac{1}{6} + (3)\frac{1}{9} + (3)\frac{1}{12} + (3)\frac{1}{15}$. The important observation here is that not only are the sums the same, but the area of the ith rectangle is the same in each case. This fact could be easily overlooked if the students calculate without looking for a pattern. By repeating this procedure with different values of a, b, and n, the students should be able to see what is happening in the general case. When $[1, b]$ is divided into n equal subintervals each of length $\Delta x = \frac{b-1}{n}$ and $[a, ab]$ is divided into n equal subintervals of length $\Delta x' = \frac{ba - a}{n}$, then $\Delta x' = a\Delta x$. If c_i is the left-hand endpoint of the ith subinterval of $[1, b]$, then it is not to hard to see that c'_i, the left-hand endpoint of the ith subinterval of

$[a, ab]$, is $c'_i = ac_i$. Thus the area of the ith rectangle for the interval $[1, b]$ is $\frac{\Delta x}{c_i}$ which equals the area of the ith rectangle for the $[a, ab]$ interval, namely $\frac{\Delta x'}{c'_i} = \frac{a\Delta x}{ac_i}$. Note that there is no reason to use the left-hand end point in this argument. If c_i is the sampling point of the ith subinterval of $[1, b]$, then choosing $c'_i = ac_i$ as the sampling point of the ith subinterval of $[a, ab]$ will insure that the area of the ith rectangle is the same in each Riemann sum.

To conclude that the areas under the curve are equal, the students must explain how the area is defined as the limit (as the mesh of the partition goes to 0) of Riemann sums. Since for each value of Δx and $\Delta x' = a\Delta x$, the Riemann sums can be chosen to be the same for both regions, the areas must be equal. Granted, there are a few more points that technically need mentioning. For example, one must note that the function is integrable over the intervals in order to know that any sequence of Riemann sums whose mesh goes to 0 will converge to the value of the integral.

Once it has been established that $\ln ab = \ln a + \ln b$ and $\ln a^{-1} = -\ln a$, it is a simple observation that $\ln a/b = \ln ab^{-1} = \ln a - \ln b$. It is somewhat more difficult to prove $\ln a^b = b \ln a$. A natural place to start would be to try to prove that $\ln a^m = m \ln a$ where m is an integer. For example, $\ln a^2 = \ln a{\cdot}a = \ln a + \ln a$. Considering $m = 2,3,4. . .$, most students should be able to come up with a convincing argument for the case where $m \geq 0$ even if they don't know how to use induction properly. If they do know induction, then they have an excellent opportunity to use it here. It follows immediately that even if the exponent is a negative integer, then $\ln a^m = m \ln a$. The next case one might consider is when the exponent is rational, that is, to prove

$$\ln a^{\frac{m}{n}} = \frac{m}{n} \ln a$$

where m and $n \neq 0$ are integers. This can be done using the properties that have already been established. Since m is an integer, $\ln a^{m/n} = m \ln a^{1/n}$. Since $n \ln a^{1/n} = \ln((a^{1/n})^n) = \ln a$ it follows that $\ln a^{1/n} = 1/n \ln a$ and hence $\ln a^{m/n} = \frac{m}{n} \ln a$.

A proof of the general exponent rule for logarithms, $\ln a^b = b \ln a$ where b is *any* real number, is probably beyond the reach of most calculus students. One way do so, if the students have gotten this far, would be to observe that since $\ln x$ and a^x are continuous functions, so is $\ln a^x$. Then one could take b_n to be a sequence of rational numbers which converge to b and then say that since $\ln a^{b_n} \to \ln a^b$ and $b_n \ln a \to b \ln a$ (as $n \to \infty$), it follows that $b \ln a = \ln a^b$.

It is fun to try to prove the exponent rule for logarithms independently of the other rules, using only Riemann sums. Partition $[1, a]$ into n subintervals using $1 = x_0 \leq x_1 \ldots \leq x_n = a$ and partition $[1, a^b]$ using $1 = x_0^b \leq x_1^b \ldots \leq x_n^b = a^b$. Again the problem is to arrange it so that the ith rectangle in the Riemann sum is the same in each case. Let c_i be the sampling point for $[x_i, x_{i+1}]$ and use c_i^b as the sampling point for $[x_i^b, x_{i+1}^b]$. Since it is hoped that

$$b\int_1^a \frac{1}{t}\,dt = \int_1^{a^b} \frac{1}{t}\,dt,$$

it would be very nice if one could choose c_i so that

$$b\frac{x_{i+1} - x_i}{c_i} = \frac{x_{i+1}^b - x_i^b}{c_i^b}.$$

This is true if and only if c_i satisfies

$$c_i = \left(\frac{x_{i+1}^b - x_i^b}{b(x_{i+1} - x_i)}\right)^{\frac{1}{b-1}}.$$

In this form it certainly will not be obvious to most calculus students that $c_i \in [x_i, x_{i+1}]$. One could look at some special cases. If $b = m = 1,2,\ldots$, then

$$c_i = \left(\frac{x_{i+1}^m - x_i^m}{m(x_{i+1} - x_i)}\right)^{\frac{1}{m-1}}$$

$$= \left(\frac{x_{i+1}^{m-1} + x_{i+1}^{m-2}x_i + \cdots x_{i+1}x_i^{m-2} + x_i^{m-1}}{m}\right)^{\frac{1}{m-1}}.$$

For $m = 2$, this becomes

$$c_i = \frac{x_{i+1} + x_i}{2}.$$

That is, c_i can be chosen as the midpoint of each interval. For other values of m it becomes harder to see that c_i is between x_i and x_{i+1}. For example, with $m = 3$ one gets

$$c_i^2 = \frac{x_{i+1}^2 + x_{i+1}x_i + x_i^2}{3}$$

which can be rewritten as

$$3 = \frac{x_{i+1}^2 + x_{i+1}x_i + x_i^2}{c_i^2}.$$

View the right-hand side as a function of c_i. Letting $c_i = x_i$, one sees that

$$3 \leq \frac{x_{i+1}^2 + x_{i+1}x_i + x_i^2}{x_i^2},$$

while for $c_i = x_{i+1}$

$$\frac{x_{i+1}^2 + x_{i+1}x_i + x_i^2}{x_{i+1}^2} \leq 3.$$

From the Intermediate Value Theorem, there must be a choice for c_i between x_i and x_{i+1} which makes the quantity exactly 3. A similar analysis works for higher values of m. If b is not a natural number, then the situation is a little stickier. One way of establishing the existence of a value $c_i \in [x_i, x_{i+1}]$ so that

$$c_i^{b-1} = \frac{x_{i+1}^b - x_i^b}{b(x_{i+1} - x_i)}$$

would be to rewrite it as

$$bc_i^{b-1} = \frac{x_{i+1}^b - x_i^b}{x_{i+1} - x_i}$$

and use the Mean Value Theorem with $y = x^b$. This argument works for any real value of b if one "knows" the derivative of x^b is bx^{b-1}.

Conjectures we expect that some students will make: This has been covered in the text of the sample solution.

Questions for further exploration: A nice project with which to follow this one would be to estimate *with error* the values of $\ln x$ for, say, $x = 2$ or $x = e$. This could be a useful project to do before a formal introduction to numerical integration techniques. Since the students are already thinking in terms of Riemann sums, the process of simply estimating the value of $\ln x$ should be quite straightforward. However, they will probably not be too familiar with the idea of an error bound. The key here is that since the integrand $1/x$ is a decreasing function, left Riemann sums will always overestimate the integral while right Riemann sums will always underestimate the integral. Thus one can use the difference between these two sums as an error estimate. This is intuitive enough for students to discover on their own. Using this approach, students should be able to get a start on deriving error estimates for the value of the integral.

Special implementation suggestions: The sketch of proof given in the problem statement is given in order to "jump start" the students. Honors students or groups which will receive a lot of instructor assistance could be given the definition and Part II only. Another option, which would give the students experience with Riemann sums, would be to ask the student to prove as many of the properties as possible independently of each other using only Riemann sums. Doing so, the students might discover the arguments similar to the ones above and at the same time gain more confidence in their ability to work with Riemann sums. A class meeting about halfway through, where the students could present preliminary results, would be helpful.

Title: Numerical Integration and Error Estimation

Author: Steve Boyce, Berea College

Problem Statement: Among the many concepts encountered in introductory calculus, the definite integral is perhaps the most difficult. Intricacies of the definition may seem unimportant once introduction of the Fundamental Theorem encourages thinking about the integral in terms of the computational short cut

$$\int_a^b f(x)dx = F(b) - F(a) \text{ where } F'(x) = f(x).$$

However, problems are soon encountered if one confuses this result with the *meaning* of the integral. For example, the integrals

$$\int_0^1 e^{x^2} dx \text{ and } \int_0^1 \sqrt{1 - 4\sin^2(x)}dx$$

cannot be evaluated using the Fundamental Theorem because neither integrand has an antiderivative which can be expressed in terms of elementary functions. This does *not* mean that the integrands have no antiderivatives; part of what the Fundamental Theorem guarantees is that every continuous function is the derivative of *some function.* What it does mean, roughly, is that you have no way to express the antiderivatives as simple combinations of well-known functions and, more importantly here, no convenient way to evaluate them at the limits of integration.

Two questions emerge from these observations. First, how can a given integral be evaluated in case an appropriate antiderivative cannot be found? Second, what assurance is there that the integral even has a value if the antiderivative cannot be found? That is, what assurance is there that the integral *exists* or, equivalently, that the function is integrable? Attempts to address the first question lead down the path of **numerical integration** in search of methods that can be used to estimate integral values to as many decimal places as desired. This is analogous to the use of Newton's Method to estimate zeros of functions. Pursuing the second question leads to **existence theorems**, statements guaranteeing the existence of the integral for certain classes of functions. In this project you will address the first question, that of numerical integration.

Part I: For the special class of functions which are both continuous and monotone on $[a, b]$:

a) Invent a numerical integration method that can be used to evaluate definite integrals to any prescribed accuracy. A central feature of your method should be a formula for calculating an upper bound on the error. Exploit the fact that you are restricting your attention to the class of continuous and *monotone* functions.

b) As a check on its precision, use your method to approximate $\int_0^{0.1} x\,dx$ to within 0.001. Determine the actual error of your method in your approximation.

c) How many additions would your method require to approximate $\int_0^1 x^5 dx$ to within 0.01? To within 0.0001? To within 0.000001?

d) By comparison, answer the three questions in part (c) using Simpson's Rule instead of your method. How do you explain the differences?

e) How could your method be used to approximate $\int_0^2 (x^3 - 2x)dx$?

Part II: Define U_n and L_n to be the upper sum (largest Riemann sum for any particular value of n) and lower sum (smallest Riemann sum for any particular value of n), respectively, and (for any particular value of n) let S_n represent any Riemann sum for a function f on $[a, b]$. Justify the following three observations:

a) $L_n \le S_n \le U_n$

b) $L_n \le \int_a^b f(x)dx \le U_n$

c) $\left| \int_a^b f(x)dx - S_n \right| \le U_n - L_n$.

You should find it helpful to use the following version of the definition of the definite integral:

Definition: Let f be a continuous function on $[a,b]$. For each positive integer n, divide $[a,b]$ into n subintervals, each with width $\frac{b-a}{n}$, such that $a = x_0 < x_1 < x_2 < \ldots < x_n = b$. In each subinterval $[x_{i-1}, x_i]$, choose a point c_i and form the Riemann sum

$$S_n = \sum_{i=1}^{n} f(c_i)(x_i - x_{i-1}) = \sum_{i=1}^{n} f(c_i)\frac{b-a}{n}.$$

If $\lim_{n\to\infty} S_n$ exists and has the same value no matter how the c_i's are chosen, then we say the integral of f

from a to b exists and is equal to $\lim_{n \to \infty} S_n$.

This particular definition of the definite integral is probably not the same as the one in your calculus text, but for continuous functions all forms of the definition that you are likely to encounter are equivalent.

Information for the instructor only:

Problem abstract: The primary goal is to involve students in experiences that suggest the fundamental importance of understanding the definite integral as the limit of a sequence of sums. Sub-goals include involving students in experiences that lead to the discovery and use of an error bound result, and an appreciation of the weakness of an error bound that is inversely proportional to n as opposed to a higher power of n.

Prerequisite skills and knowledge: This project can be assigned any time after students have been introduced to the definition of the definite integral. Some exposure to numerical integration and error bounds would be helpful, but it is not essential. Students not acquainted with Simpson's Rule will need consult the literature to find an expression for its error bound and examples of its use.

Essential/useful library resources: David Smith's paperback *Interface: Calculus and the Computer*, (Saunders, 1984) is a wonderful source of projects related to numerical integration.

Essential/useful computational resources: none

Example of an acceptable approach:

Part I

a) Students may use any number of methods. Most likely are upper sum, lower sum and midpoint.

b) Since $f(x) = x$ and $[a, b] = [0, 0.1]$,

$$|f(b)-f(a)|\frac{b-a}{n} = \frac{(0.1)(0.1)}{n} \leq 0.001 \text{ when } n \geq \frac{(0.1)(0.1)}{0.001} = 10.$$

Using the upper sum as an approximator with $n = 10$ yields

$$(f(x_1)+f(x_2)+\ldots+f(x_n))\frac{b-a}{n} = (\frac{1}{100}+\frac{2}{100}+\ldots+\frac{10}{100})\frac{1}{100} = 0.0055.$$

Since the Fundamental Theorem can be used to see that the integral's true value is 0.005, the actual error in this approximation is 0.0005 – less than the prescribed 0.001.

c) Since $f(x) = x^5$ and $[a, b] = [0, 1]$,

$$|f(b)-f(a)|\frac{b-a}{n} = \frac{1}{n} \leq E \text{ when } n \geq \frac{1}{E}.$$

It follows that when the error tolerance "E" is specified as 0.01, 0.0001 or 0.000001, the corresponding required values for n are 100, 10,000 and 1,000,000 respectively.

d) The error bound for Simpson's Rule can be used in similar fashion to establish a linkage between the error tolerance "E" and the required number of terms, n:

$$\frac{(b-a)^5 \max\left|f^{(4)}(x)\right|}{180n^4} = \frac{120}{180n^4} \le E \text{ when } n \ge \sqrt[4]{\frac{120}{180E}}.$$

Corresponding to the "E" values 0.01, 0.0001 and 0.000001 are n values (rounded-up integer values) of 3, 10 and 29.

Why the big difference? The principal reason is the difference between n and n^4 in the two error bounds. One error bound is proportional to the fourth power of the subinterval widths while the other is proportional to their first power. It follows that as n increases and the subinterval widths become small, one error bound (the one for Simpson's Rule) decreases much more rapidly than the other.

e) The problem here is that $f(x) = x^2 - 2x$ is not monotone on [0,2]. However, f is decreasing on [0, 1] and increasing on [1, 2]; that is, the function is monotone on each of two subintervals whose union is [0, 2]. So the problem can be overcome by expressing the original integral as the sum two integrals, one from 0 to 1 and the other from 1 to 2.

Part II

There are two key insights students need to achieve, with or without instructor assistance. One is that observation c) means that, for a given partition, the difference between the upper and lower sums provides an upper bound on the error involved in approximating the integral by the value of any Riemann sum associated with that particular partition. Second, limiting consideration to monotone functions leads to a very simple expression for this error bound. If, for example, f is increasing on $[a, b]$, then the upper and lower sums are easy to form since the maximums of f occur at the right endpoint of each subinterval and its minimum values at the left endpoint:

$$U_n = (f(x_1) + f(x_2) + \ldots + f(x_n))\frac{b-a}{n} \text{ and } L_n = (f(x_0) + f(x_1) + \ldots + f(x_{n-1}))\frac{b-a}{n}.$$

As a result, all but two terms cancel when the lower sum is subtracted from the upper sum: $U_n - L_n = (f(x_n) - f(x_0))\frac{b-a}{n} = (f(b) - f(a))\frac{b-a}{n}$. This expression changes only in sign if f is decreasing on $[a, b]$, so in either case we obtain $U_n - L_n = |f(b) - f(a)|\frac{b-a}{n}$.

Questions for further exploration: (1) A very pleasing extension of these ideas can be seen in the way approximations from the midpoint and trapezoidal rules (as opposed to the upper and lower sums) can be combined to establish an error bound for functions whose concavity does not change on $[a, b]$. See the Smith book mentioned above, Chapter 26. (2) The project following this one would give students a chance to use the strategy of assuming a special hypothesis (that the functions are monotone in this case) as part of a conjecture-proof investigation. The central goal of that project is to lead students to think more carefully about what the integral definition actually says and what must be accomplished to show that the definition is satisfied.

References/bibliography/related topics: *Interface: Calculus and the Computer* by David Smith (Saunders, 1984)

Special implementation suggestions: For this project it seems particularly important to schedule a progress report soon after the student group has begun its work. If the group seems not to be on a productive track, it may be enough to point out the significance of II(c), and with that motivation suggest they examine $U_n - L_n$ in hopes that something useful will emerge. If there is confusion about what "useful" means in this context (i.e., about the meaning error bound), consulting the literature for the statement and use of the error bound for Simpson's Rule — a job which needs to be done eventually anyway — should be helpful.

Since II(c) can now be seen to mean that

$$\left|\int_a^b f(x)dx - S_n\right| \leq |f(b) - f(a)|\frac{b-a}{n},$$

it follows that any Riemann sum can be used to approximate the integral as long as n is large enough to make the error bound at least as small as the specified error tolerance. Students seem most likely to choose either the upper or the lower sum; it would be especially pleasing to have a group suggest using the average of those two.

Title: An Integral Existence Theorem

Author: Steve Boyce, Berea College

Problem Statement: Among the many concepts encountered in introductory calculus, the definite integral is perhaps the most difficult. Intricacies of the definition may seem unimportant once introduction of the Fundamental Theorem encourages thinking about the integral in terms of the computational short cut

$$\int_a^b f(x)dx = F(b) - F(a) \text{ where } F'(x) = f(x).$$

However, problems are soon encountered if one confuses this result with the *meaning* of the integral. For example, the integrals

$$\int_0^1 e^{x^2}\,dx \text{ and } \int_0^1 \sqrt{1-4\sin^2(x)}dx$$

cannot be evaluated easily using the Fundamental Theorem because neither integrand has an antiderivative which can be expressed in terms of elementary functions. This does *not* mean that the integrands have no antiderivatives; part of what the Fundamental Theorem guarantees is that every continuous function is the derivative of something. What it does mean, roughly, is that you have no way to express the antiderivatives as simple combinations of well-known functions and, more importantly here, no convenient way to evaluate them at the limits of integration.

Two questions emerge from these observations. First, how can a given integral be evaluated in case the appropriate antiderivative cannot be found? Second, what assurance is there that the integral even has a value if the antiderivative cannot be found? That is, what assurance is there that the integral exists or, equivalently, that the function is integrable? Attempts to address the first question lead down the path of **numerical integration** in search of methods that can be used to estimate integral values to as many decimal places as desired. Pursuing the second question leads to **existence theorems**, statements guaranteeing the existence of the integral for certain classes of functions. For example, an existence theorem usually stated (without proof) in introductory calculus texts is this:

Theorem: If f is continuous on the closed interval $[a, b]$, then $\int_a^b f(x)\,dx$ exists.

In this project you will prove the integral existence theorem for the special class of functions which are both continuous and *monotone* on $[a, b]$. This task has a theoretical focus in which the definition of definite integral plays a starring role. You should find it helpful to use the following version of the definition.

Definition: Let f be a continuous function on $[a, b]$. For each positive integer n, divide $[a, b]$ into n equal-width subintervals, $[x_{i-1}, x_i]$, such that $a = x_0 < x_1 < x_2 < ... < x_n = b$. In each subinterval $[x_{i-1}, x_i]$, choose a point c_i and form the Riemann sum

$$S_n = \sum_{i=1}^{n} f(c_i)(x_i - x_{i-1}) = \sum_{i=1}^{n} f(c_i)\frac{b-a}{n}.$$

If $\lim_{n\to\infty} S_n$ exists and has the same value no matter how the c_i's are chosen, then we say the integral of f from a to b exists and is equal to $\lim_{n\to\infty} S_n$.

This definition of the definite integral is probably not the same as the one in your calculus text, but for continuous functions all forms of the definition that you are likely to encounter are equivalent. Also, it will be helpful to you if you know how to apply what is often called the "Squeeze" or the "Pinching" Theorem.

Part I: Define U_n and L_n to be the upper sum (largest Riemann sum for any particular value of n) and lower sum (smallest Riemann sum for any particular value of n), respectively. Let S_n represent any Riemann sum for a function f on $[a, b]$. Verify the following:

a) $L_n \le S_n \le U_n$

b) $L_n \le \int_a^b f(x)\,dx \le U_n$

c) $\left|\int_a^b f(x)dx - S_n\right| \le U_n - L_n.$

Remember that the function f you are dealing with is monotone!

Part II: For the special class of functions which are both continuous and monotone on $[a, b]$, prove the integral existence theorem.

Information for the instructor only:

Problem abstract: The primary goals are to involve students in experiences that suggest the fundamental importance of understanding the definite integral as the limit of a sequence of sums and illustrate the strategy of assuming a special hypothesis (that the functions are monotone in this case) as part of a conjecture-proof investigation. The central goal is to lead students to think more carefully about what the integral definition actually says and what must be accomplished to show that the definition is satisfied.

Prerequisite skills and knowledge: This project can be assigned any time after students have been introduced to the definition of the definite integral.

Essential/useful library resources David Smith's paperback *Interface: Calculus and the Computer* (second edition; Saunders, 1984) is a wonderful source of projects related to numerical integration. Chapters 24 and 25 contain material related to this project.

Essential/useful computational resources: none

Example of an acceptable approach: Because of their lack of experience with proof, students will almost certainly have difficulty translating the definition of definite integral into a tentative plan for showing that the definition is satisfied in the case of monotone functions; that is, in first identifying a statement that begins, "It would be enough to show . . . ." If initial student efforts produce nothing but frustration, it should be helpful to lead them to the point of understanding that it would be enough to show the following: "If $\{S_n\}$ and $\{T_n\}$ are any two sequences of Riemann sums, where n represents the number of subintervals, then the limits of $\{S_n\}$ and $\{T_n\}$ as n approaches infinity both exist and have the same value."

Once this plan is in place, observation I a) suggests a very geometric, visual explanation as to why S_n and T_n are forced arbitrarily close together as n increases. It implies that both S_n and T_n are in the closed interval $[L_n, U_n]$. That is, the distance between S_n and T_n is no greater than $U_n - L_n$, so there is motivation to examine what special form this difference might have in the case of monotone functions. Alternatively, this point might be reached from observation (1) by an inequality argument:

$$L_n \leq S_n \leq U_n \text{ and } -U_n \leq -T_n \leq -L_n$$
$$\Rightarrow -(U_n - L_n) \leq (S_n - T_n) \leq (U_n - L_n)$$
$$\Rightarrow 0 \leq |S_n - T_n| \leq (U_n - L_n).$$

The argument can be concluded by observing that for monotone functions,

$$U_n - L_n = |f(b) - f(a)| \frac{b-a}{n},$$

and citing the Squeeze Theorem.

References/bibliography/related topics: *Interface: Calculus and the Computer* by David Smith (second edition, Saunders, 1984).

Special implementation suggestions: This problem could be assigned to students who had successfully completed the project preceding this one. It could be done by individuals instead of small groups.

Special evaluation suggestions: You will need to be careful in how you judge students' success on this project because few will have had previous experience in writing good mathematical proofs. You may want to allow them to correct and resubmit their proofs after you have first evaluated them.

Title: A Fundamental Project

Author: Charles Jones, Grinnell College

Problem Statement: This project deals with "area functions" defined as follows: Given a function f, define a new (area) function F by $F(x)$ = area bounded by the t-axis (horizontal axis), the vertical lines t = constant and $t = x$, and the graph of $y = f(t)$. (See Figure 1.)

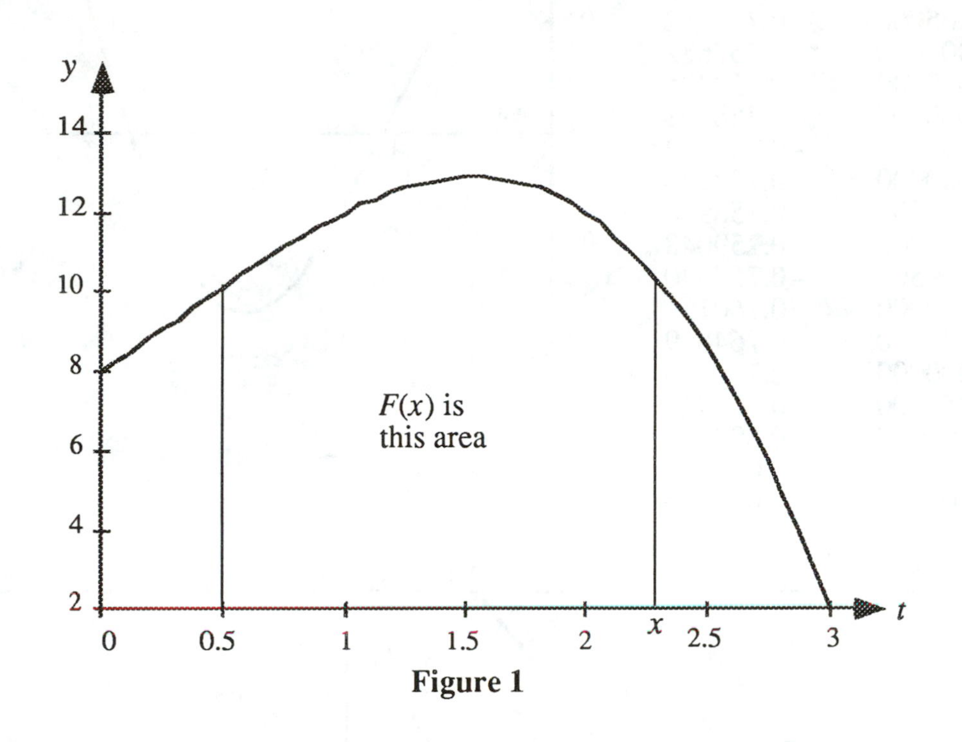

Figure 1

Problem 1: Suppose $f(t) = 2t + 3$. Define $F(x)$ to be the area bounded by $t = 1$, $t = x$, the t-axis and $y = f(t)$.

(a) Graph $y = f(t)$ and compute $F(4)$.

(b) Compute $F(1)$.

(c) Find a general formula for $F(x)$, if $x \geq 1$.

This project also deals with derivatives. In particular, you will need to compute, approximately, the derivative of a function, call it g, by calculating the difference quotient,

$$\frac{g(x + \Delta x) - g(x)}{\Delta x},$$

for various small values of Δx (both positive and negative). By noting the values of the difference quotient for values of Δx near 0, you will make an educated guess of the value of the derivative.

Problem 2: Refer to Figure 2 for a table of values for a function g and graphs of $y = g(x)$ on three different domains. Sketch the tangent line to the graph at $x = 2$, and compute the derivative of $g(x)$ there by the above method. Does your educated guess agree with the slope?

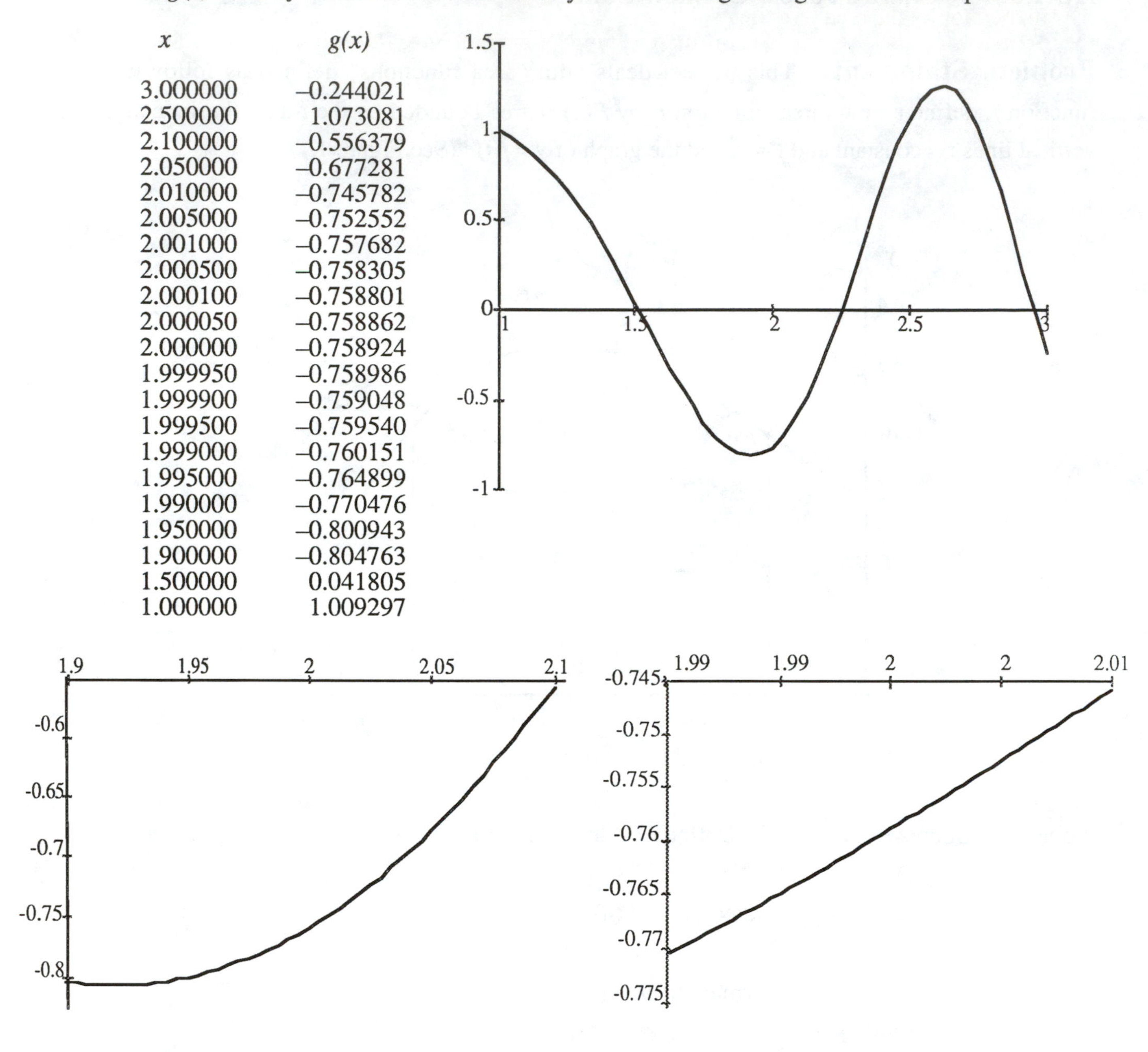

x	$g(x)$
3.000000	−0.244021
2.500000	1.073081
2.100000	−0.556379
2.050000	−0.677281
2.010000	−0.745782
2.005000	−0.752552
2.001000	−0.757682
2.000500	−0.758305
2.000100	−0.758801
2.000050	−0.758862
2.000000	−0.758924
1.999950	−0.758986
1.999900	−0.759048
1.999500	−0.759540
1.999000	−0.760151
1.995000	−0.764899
1.990000	−0.770476
1.950000	−0.800943
1.900000	−0.804763
1.500000	0.041805
1.000000	1.009297

Figure 2

The following "fundamental problems" tie together the concepts of area functions and derivatives.

Problem 3: Refer to Figure 3 for a table of values of a function f and graphs of $y = f(t)$ on three different domains. Define $F(x)$ to be the area bounded by $t = 1$, $t = x$, the t-axis, and the graph of $y = f(t)$.

(a) Sketch a region whose area is $F(3)$.

(b) Write the difference quotient for $F'(3)$.

(c) Sketch the region whose area is represented by the numerator of the difference quotient if $\Delta x = 0.5$.

(d) Approximate the value of the difference quotient for $F'(3)$ when $\Delta x = 0.5$ by approximating the numerator by the area of a trapezoid.

(e) Approximate the values of the difference quotient for several small values (positive and negative) of Δx. Note: When Δx is negative, the value of the difference quotient numerator will be the negative of an area.

(f) Use your approximations to guess the value of $F'(3)$. How is this related to f?

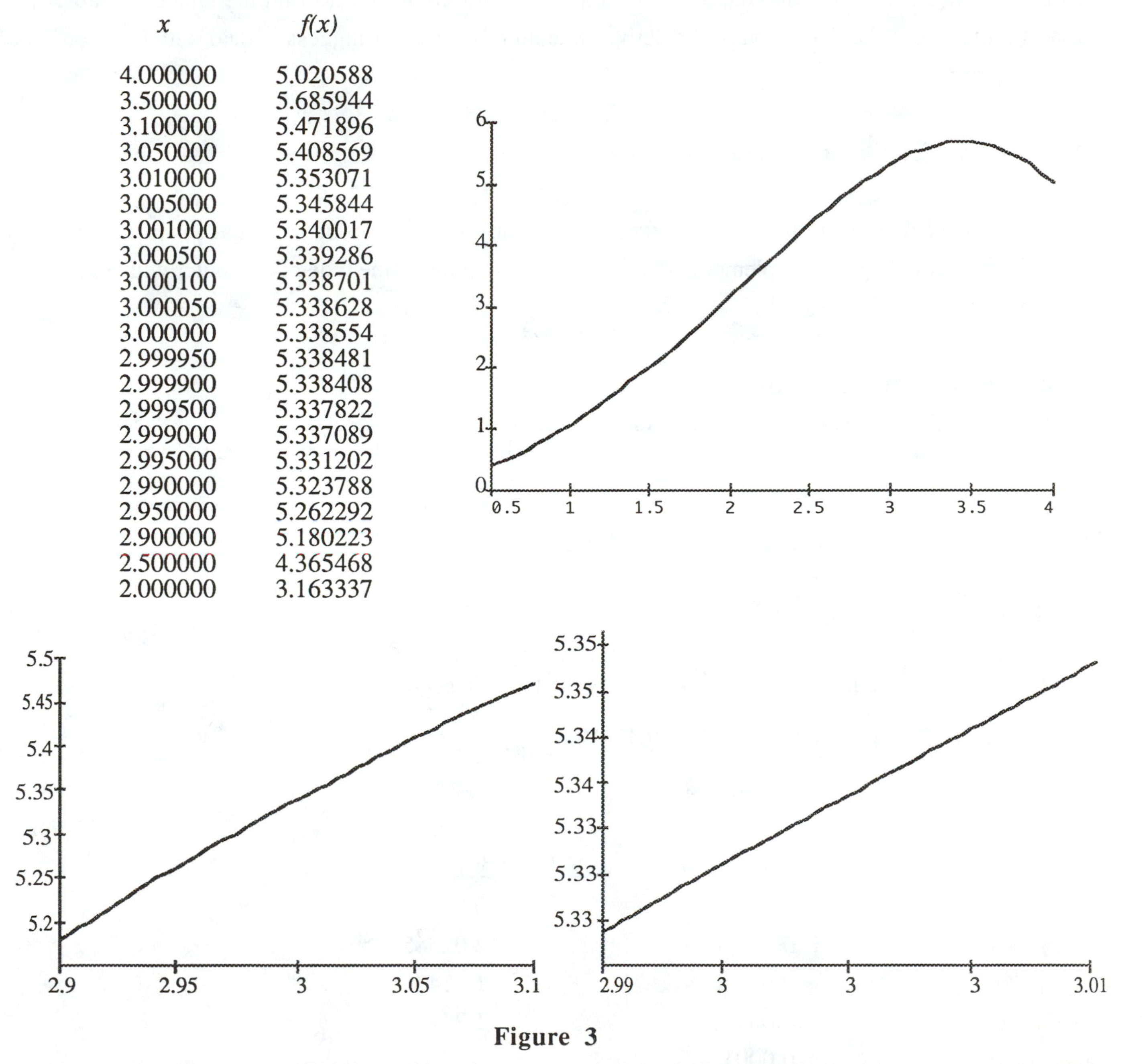

x	$f(x)$
4.000000	5.020588
3.500000	5.685944
3.100000	5.471896
3.050000	5.408569
3.010000	5.353071
3.005000	5.345844
3.001000	5.340017
3.000500	5.339286
3.000100	5.338701
3.000050	5.338628
3.000000	5.338554
2.999950	5.338481
2.999900	5.338408
2.999500	5.337822
2.999000	5.337089
2.995000	5.331202
2.990000	5.323788
2.950000	5.262292
2.900000	5.180223
2.500000	4.365468
2.000000	3.163337

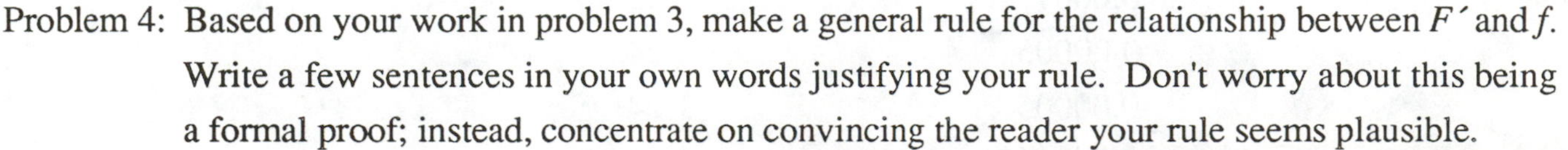

Figure 3

Problem 4: Based on your work in problem 3, make a general rule for the relationship between F' and f. Write a few sentences in your own words justifying your rule. Don't worry about this being a formal proof; instead, concentrate on convincing the reader your rule seems plausible.

Information for the instructor only:

Problem abstract: The goal of this project is to have students discover the Fundamental Theorem of Calculus and explain why their version is a plausible result. This project can be assigned any time after the definition of the derivative, and it should be completed before the Fundamental Theorem is covered in class.

Prerequisite skills and knowledge: The students should know how to find the area of a trapezoid, have ability to use the definition of the derivative, and possess a willingness to deal with functions not given by formulas.

Essential/useful library resources: none

Essential/useful computational resources: None are needed for the original project; however, use of the section "Special implementation suggestions" requires some type of computer with a program to compute function values.

Example(s) of an acceptable approach:

1. $f(t) = 2t + 3$. $F(x)$ = area bounded by $t = 1$, $t = x$, the t-axis, and $y = f(t)$.
   - (a) $x = 4, F(3) = 16 + 12 - 4 = 24$
   - (b) $x = 1, F(1) = 0$
   - (c) $\text{Area} = \text{width}\left[\dfrac{\text{height 1 + height 2}}{2}\right] = (x-1)\left[\dfrac{(5+2x)+3}{2}\right]$
     $= (x - 1)(4 + x) = x^2 + 3x - 4.$

2. The function used in this problem is given by the formula $g(x) = \dfrac{x}{10} + \sin(x^2 + 1)$.

   So, $g'(2) = \dfrac{1}{10} + 4\cos(5) \approx 1.234647742$.

   A sample solution follows:

Δx	Value of difference quotient
1.0	0.514903
0.1	2.02545
0.01	1.3142
0.001	1.242
0.0001	1.23
0.00005	1.24
–0.00005	1.24
–0.0001	1.24

3. (b) Difference quotient $= \dfrac{\text{Area bounded by } t = 3,\ t = 3 + \Delta x, y = f(t), t-\text{axis}}{\Delta t}$

(d) $\dfrac{\dfrac{5.685944 + 5.338554}{2}(0.5)}{0.5} = 5.512249$

(e)

Δx	Difference quotient
0.1	5.405225
0.01	5.3458125
0.001	5.3392855
0.0001	5.3386275
0.00005	5.338591
–0.00005	5.3385175
–0.0001	5.338481

(f) Guess: The average of the difference quotient when $\Delta x = 0.00005$ and $\Delta x = -0.00005$ is 5.33855425. This appears to be $f(3)$!

4. When you compute the difference quotient F, the numerator is the average of the heights, $f(x)$ and $f(x + \Delta x)$. You then multiply by the width, Δx, to get the area of the trapezoid. However, the denominator of the difference quotient is also Δx; so, the quotient is just the average height, $\dfrac{f(x) + f(x + \Delta x)}{2}$. Since $f(x + \Delta x)$ gets close to $f(x)$ as $\Delta x \to 0$, this average height goes to $f(x)$. Thus, $F'(x) = f(x)$.

   This is the type of argument I would hope to see. Note that no mention is made of the continuity of f, and that students will consider "$f(x + \Delta x)$ gets close to $f(x)$" reasonable.

Questions for further exploration: Modify the functions given to have negative values.

Special implementation suggestions: Instead of dealing with a table of function values, an interactive computer program to serve as a "black-box" function would work nicely.

APPLICATIONS OF INTEGRATION

This section includes those projects which are relevant to the material covered in the middle third of a standard calculus sequence. Actually, the first project, *Inventory Decisions*, is a one variable max/min problem and could probably be done much sooner by honors-level students or as a classroom project. It is put in this section as it is probably more accessible to most students after they have some familiarity with the average value of a function. There is a wide variety of applications of integration in this set of projects. Students can have the opportunity to see an artistic application in *Tile Design* (an application quite different from those typically encountered in calculus), a challenging but more standard geometric application in *The Ice Cream Cone Problem*, or an application to a mathematical question in *Minimizing the Area Between a Graph and Its Tangent Lines*. Further applications are found in *Riemann Sums, Integrals, and Average Values* where students are given practice in deriving integral formulas for applications by taking limits of Riemann sums. Two projects from the previous section are also appropriate at this point in the course. *Population Growth* is a nice application of the derivative of the natural exponential function and *Drug Dosage* can be used as an introduction to series.

Title: Inventory Decisions

Author: Steve Boyce, Berea College

Problem Statement: A computer services firm regularly uses many cartons of computer paper. They purchase the cartons in quantity from a discount supplier in another city at a cost of $22.46 per carton, store them in a rented warehouse near company grounds and use the paper gradually as needed. There is some confusion among company managers as to how often and in what quantity paper should be ordered. On one hand, since the supplier is providing out-of-town delivery by truck, there is a basic $360 charge for every order regardless of the number of cartons purchased, assuming the order is for no more than 3,000 cartons (the truck's capacity). This cost has been used by some managers as an argument for placing large orders as infrequently as possible. On the other hand, as other managers have argued, large orders lead to large warehouse inventories and associated costs of at least two kinds that should be considered. First, they claim, whatever money is used to pay for paper that will only sit in the warehouse for a long time could instead, for a while at least, be allocated to some profit producing activity. At the very least such money could be accumulating interest in a bank account. This loss of investment opportunity and associated earnings is referred to as the "opportunity cost" resulting from the investment in paper inventory. Secondly, the company has to pay rent for the warehouse. While other company property is stored there as well, the managers agree that a fraction of the rent equal to the fraction of the warehouse space occupied by paper should be viewed as part of the cost of storing paper. These latter two costs, collectively referred to as the inventory "holding cost" and estimated to be 18 cents/carton/week, have been used to justify claims by some that paper orders should be smaller and placed more frequently. In hopes of resolving the confusion, the managers have hired you as a consultant.

After talking more with various company personnel, asking many questions and inspecting company records, you have accumulated the following summary notes. Use them along with additional modeling and analysis as the basis for a report to the managers recommending in what quantity and how often paper should be ordered.

Note 1: Company data on the number of cartons used per week is shown below. No one seems to think usage rate will change in any significant way.

Week	No.	Week	No.	Week	No.	Week	No.	Week	No.	Week	No.
1	150	10	152	19	149	28	150	37	151	46	152
2	149	11	150	20	150	29	147	38	148	47	151
3	150	12	149	21	150	30	152	39	150	48	148
4	151	13	149	22	150	31	150	40	150	49	150
5	153	14	150	23	150	32	151	41	149	50	150
6	150	15	150	24	152	33	148	42	152	51	151
7	148	16	150	25	150	34	151	43	150	52	147
8	150	17	151	26	148	35	150	44	149		
9	150	18	150	27	151	36	150	45	147		

Note 2: We should not let the paper supply run out. Managers agree that a work stoppage would be disastrous for customer relations, so paper would be purchased from a local source rather than allowing a stoppage to occur. The best local price is \$46.90 per carton compared to \$22.46 from the usual discount supplier.

Note 3: The discount supplier is *very* reliable about providing quick delivery. When an order is placed in the morning, she has never failed to deliver before 5 P.M. the same day. It seems safe to count on this. So in modeling, for simplicity, we can assume that the new order arrives just as the last stored carton is used.

Note 4: Managers seem to agree that the goal in deciding how much and how often to order should be to minimize average weekly cost associated with the purchase and storage of paper. Average weekly cost has three constituents: purchase cost (\$22.46 × number of cartons ordered/week), delivery cost (\$360 × number of orders/week), and holding cost. When pushed to be more precise about holding cost, manager consensus was that average holding cost/week should be measured as 18 cents/carton/week × the average inventory between orders (i.e., the average number of cartons stored in the warehouse from the time one order arrives to the time the next order arrives).

Note 5: It probably will simplify modeling and analysis to assume that the inventory level (the number of boxes stored) and time are continuous rather than discrete variables. That is, instead of assuming the time and inventory variables can only take on integer values representing weeks and boxes, assume they can take on any non-negative real number values. That way, if the usage rate is assumed to be constant (seems reasonable in view of the data in Note 1 above), the inventory level can be modeled as piecewise linear, as indicated in the sketch:

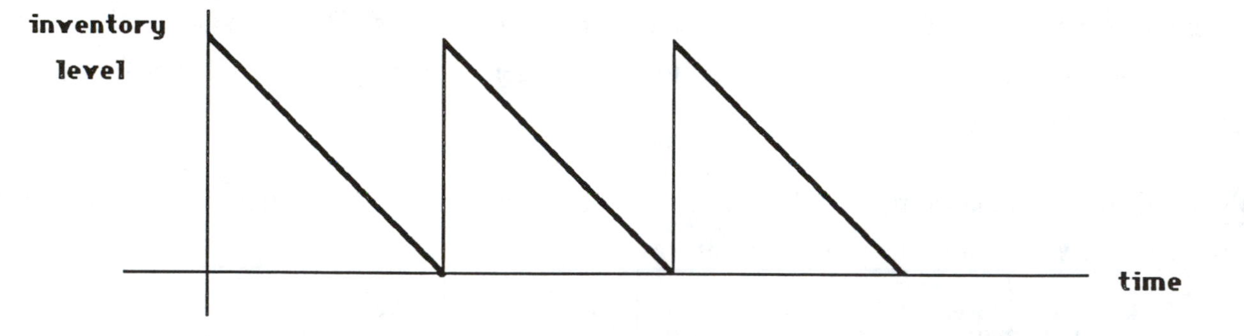

Information for the instructor only:

Problem abstract: The goals of this project are to involve students in (a) developing and using a simple but fundamental "lot size" model that appears often in inventory theory; (b) applying max/min analysis in a relatively realistic setting; (c) using graphs to investigate the properties of the model; (d) the clear and carefully reasoned written/oral explanation of project work; (e) using the model, once developed, to investigate post-optimality questions. In addressing one post-optimality question given in "Questions for further exploration," a clear advantage can be seen in using derivative analysis to locate an absolute minimum as opposed to simply estimating the location of the minimum using graphics software or a graphing calculator.

Prerequisite skills and knowledge: This project can be assigned any time after students have been introduced to the use of the derivative in max/min analysis. It is recommended, but not absolutely necessary, that it be given to students after they understand the use of the integral to find the average value of a continuous function.

Essential/useful library resources: none

Essential/useful computations resources: It would be helpful if students can use a computer or calculator to generate 2-dimensional graphs.

Example of an acceptable approach: As suggested in Note 5 of the problem statement, the modeling assumptions imply that inventory level L is piecewise linear. This is indicated in Figure 1 where R represents the constant paper usage rate in cartons/week, Q the amount to be ordered and t the time in weeks.

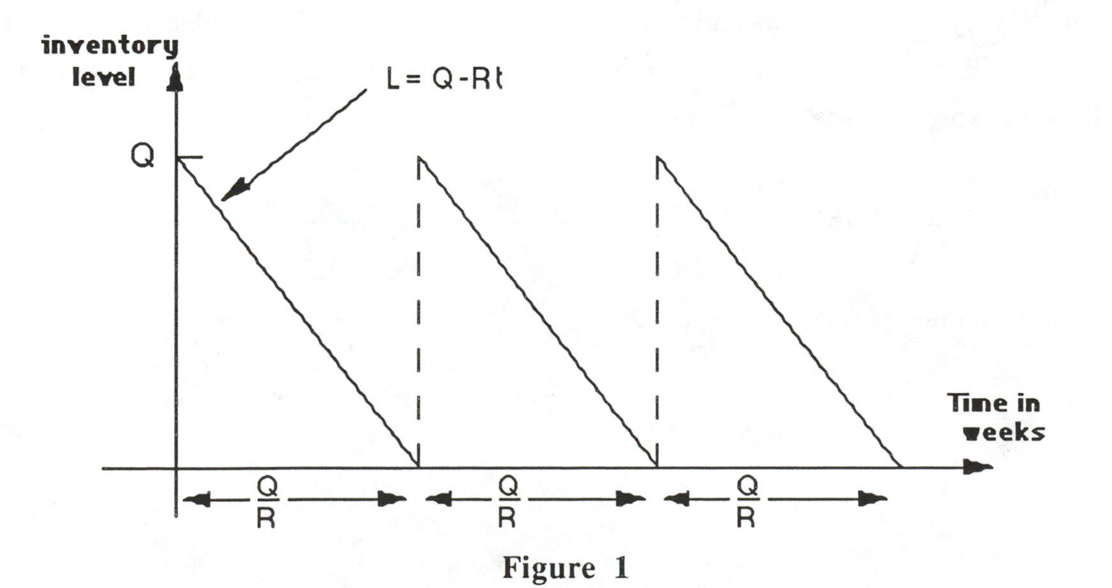

Figure 1

It is apparent from the figure that how much to order and how often to order are not independent decisions. Ordering Q implies that orders should be placed every Q/R weeks. Our task is to find the best value for Q; that is, we are to find the value of Q which minimizes A, the average weekly cost of purchasing and storing paper. The key step in doing so, and the primary modeling task left to the students, is to formulate a functional expression for the average weekly cost. The suggestion in Note 4 is to view A as the sum of three parts: A = purchase cost/week + delivery cost/week + holding cost/week. If students have difficulty with this, one suggestion which might be useful without revealing all would be to visualize A as (the average cost/cycle) $\times$ (the number of cycles per week) where "cycle" means the time, Q/R, between order arrivals. The purchase cost and delivery cost per cycle are clearly $22.46Q$ and 360, respectively. The average holding cost per *week*, according to Note 4, should be viewed as $0.18Q/2$ since the average value of the inventory over a cycle is given by

$$\frac{R}{Q}\int_0^{\frac{Q}{R}}(Q-Rt)dt=\frac{R}{Q}\left(Qt-\frac{Rt^2}{2}\right)\Bigg|_0^{\frac{Q}{R}}=Q-\frac{Q}{2}=\frac{Q}{2}.$$

The average holding cost per *cycle*, then, is $(0.18Q/2)(Q/R) = 0.18Q^2/2R$ since each cycle contains Q/R weeks.

Students not acquainted with the integral may have to argue (or be persuaded) that from the graph, it is intuitively reasonable to accept $(Q + 0)/2$ as the average value of inventory. For those acquainted with the integral as the limit of a sum, but not with its use in averaging continuous variables, this would be a good occasion for consulting the literature or, probably with instructor assistance, developing a derivation. In either case, it should be interesting to see how students respond, prior to prompting, to the necessity of finding the average value of a continuous variable.

To complete the formulation, students need to exploit the connection between period and frequency; that is, they need to understand that one order arriving every Q/R weeks is equivalent to R/Q arrivals per week. Given this, and the fact that notes 1 and 5 make it reasonable to assume R = 150, we have the following expression for average weekly cost:

$$A(Q)=\left(22.46Q+360+\frac{0.18Q^2}{2R}\right)\times\left(\frac{R}{Q}\right)=3{,}369+\frac{54{,}000}{Q}+0.09Q.$$

The graph of this function is shown in Figure 2.

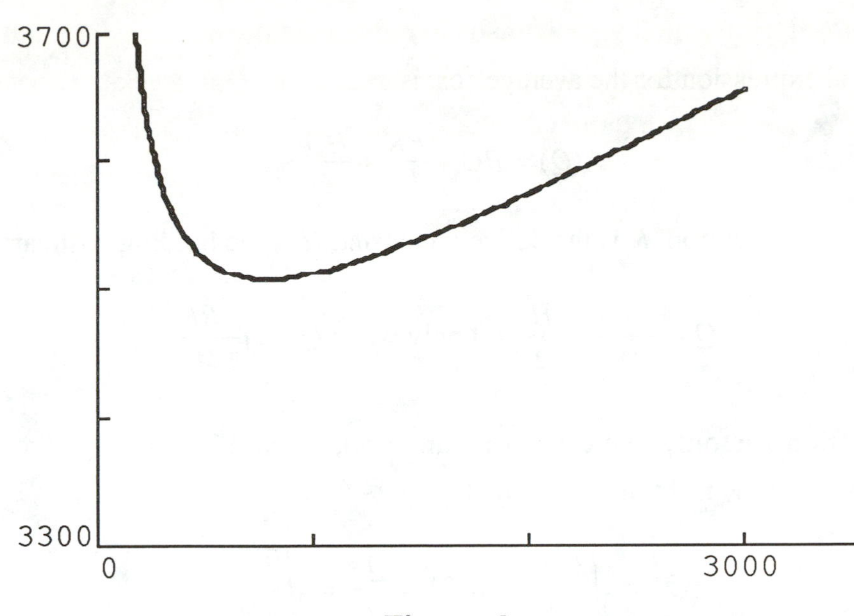

Figure 2

All that remains, except for the questions raised in notes 6 and 7, is to locate the absolute minimum of $A(Q)$ on (0,3000]. (It should be clear to students why it is not necessary to consider Q values larger than 3000.) From the graph it is clear that the minimum we seek must lie at the derivative's only zero. Since

$$A'(Q) = \frac{-54{,}000}{Q^2} + 0.09 = 0 \text{ when } Q = \sqrt{\frac{54{,}000}{0.09}} \approx 775,$$

it follows that the best amount to order, Q^*, and the best time spread between orders, T^*, are given by $Q^* \approx 775$ cartons and $T^* = Q^*/R \approx 5.16$ weeks.

Questions for further exploration:

Question 1: The supplier is considering increasing the delivery charge. How sensitive are the questions we are examining to changes in this parameter? That is, if the delivery charge is increased by a factor of p for some $p > 1$ — so that the new charge is $360p$ — how will the optimal (best) decisions regarding when and how much to order change in response?

Question 2: The supplier is considering a new policy that would provide quantity discounts. Instead of charging a flat $22.46/carton, they may begin charging $22.46/carton if less than 500 cartons are ordered, $21.96 if the order falls between 500 and 1500 and $21.46 from 1500 to 3000. If they do institute this change, would it make any difference in the optimal decisions regarding how much and when to order? How would average weekly cost be affected?

A more general expression for the average cost is

$$A(Q) = RC + \frac{RK}{Q} + \frac{HQ}{2}$$

where C is the purchase cost/carton, K is the delivery cost and H is the holding cost/carton/week. Since

$$A'Q = \frac{-RK}{Q^2} + \frac{H}{2} = 0 \text{ only when } Q = \sqrt{\frac{2RK}{H}},$$

the average cost function has only one critical point. And, since $A'(Q)$ is negative to the left of the critical point and positive to the right, it follows that

$$Q^* = \sqrt{\frac{2RK}{H}} \text{ and } T^* = \frac{Q^*}{R} = \sqrt{\frac{2K}{RH}}.$$

It is now clear that if K is replaced by pK, then both Q^* and T^* will increase by a factor equal the square root of p. This kind of insight is extremely unlikely to develop, short of some Kepler-like inductive effort, if the cost minimum is estimated numerically one example at a time.

One effective way to investigate question 2 is to graph the three average cost functions associated with the three prices being considered by the supplier. Graphs of A_1, A_2 and A_3 are shown in Figure 3 where

$$A_1(Q) = 3{,}369 + \frac{54{,}000}{Q} + .09Q \text{ if } 0 \leq Q < 500$$

$$A_2(Q) = 3{,}294 + \frac{54{,}000}{Q} + .09Q \text{ if } 500 \leq Q < 1500$$

$$A_3(Q) = 3{,}219 + \frac{54{,}000}{Q} + .09Q \text{ if } 1500 \leq Q \leq 3000.$$

Since the three differ only in the term representing average purchase cost per week — a term which does not depend on Q — their absolute minima will all occur at the same value of Q. The three graphs and the analysis carried out above for the \$22.46 case make it clear that the best values of Q for A_1, A_2 and A_3 are 500, 775 and 1500 respectively.

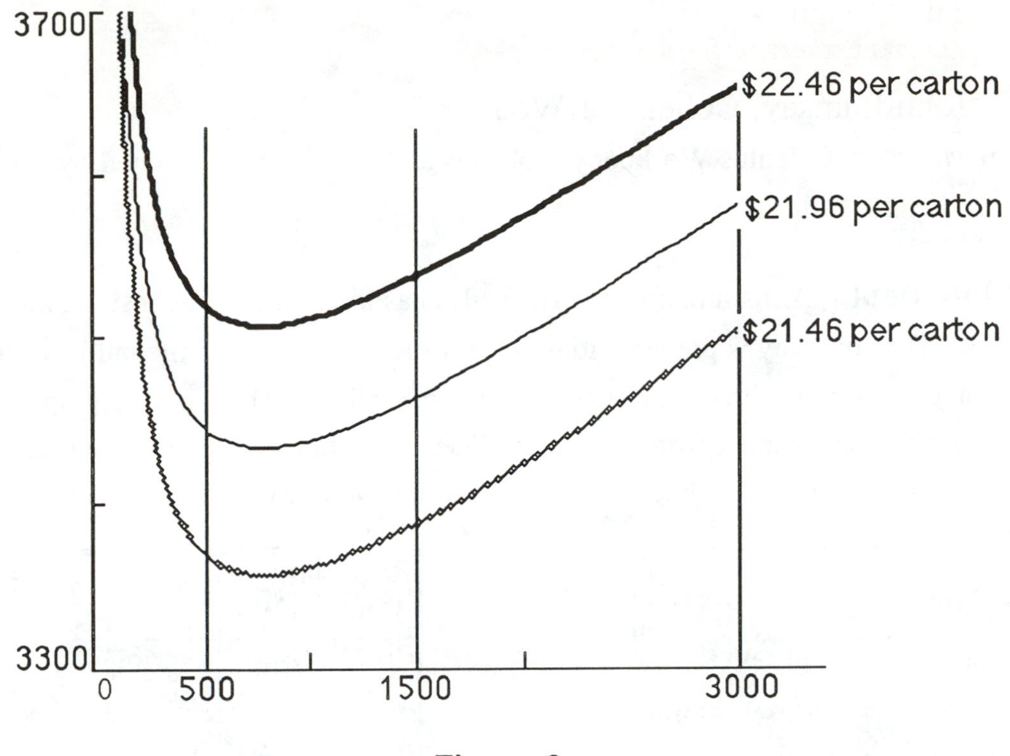

Figure 3

Also clear from the graph is the fact that of the three associated average weekly costs, $A_3(1500)$ is the least. It follows that the new Q^* is 1500 cartons and the new T^* is $Q^*/R = 1500/150 = 10$ weeks. While Q^* and T^* almost double, the change in average weekly cost from $A_2(775)$ to $A_3(1500)$ represents a little less than 1.3% improvement.

Special implementation suggestions: The instructor is *strongly* encouraged to include the two questions for further exploration in the assignment if at all possible. They are not included in the problem statement here only because they will likely require extra effort and guidance beyond that of most problems in this volume.

Title: Tile Design

Author: John Ramsay, College of Wooster
(adapted from MAPLE: Calculus Workbook Problems and Solutions, distributed by the University of Waterloo)

Problem Statement: A manufacturer of floor tiles has hired you as a consultant to design a new square floor tile. The company is prepared to provide for two colors in the tile but lacks the equipment necessary to have more than three simple regions for the colors. Through the grapevine, you have learned that the manufacturer likes symmetry and dislikes linear patterns. Following is an example of an acceptable tile design and corresponding floor pattern given to you by the manufacturer.

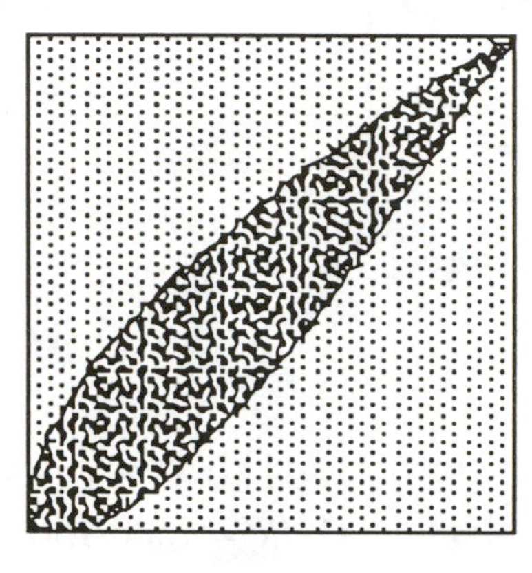

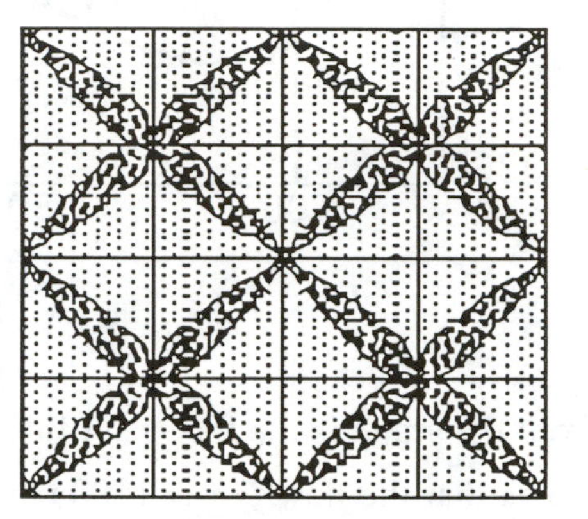

First, the manufacturer would like to know the relative amount of the two colors (percentage of tile devoted to each color) in the given example so that cost estimates (which depend on how much of each color) can be made. Model the given tile with two functions in order to estimate the areas. You must justify your approach fully. Second, use a graphing utility to design three other tile patterns which result in "pleasing" floor patterns. (For example, you might want to produce a design which yields a floor pattern of intersecting circles.) Submit a written presentation of your designs and the corresponding floor patterns for review by the manufacturer. You should also indicate the relative amounts of the two colors for each of your patterns.

Information for the instructor only:

Problem abstract: This is a project which allows for a great deal of creativity on the part of the students and combines mathematics with art in a way that most calculus students do not have the opportunity to see. As a result of the mathematical, artistic and communication (oral and/or written) skills required in the project, it will provide an opportunity for all students in the group to contribute in a significant way. The project could be given to one group of students or could be a class project with several groups presenting "rival" proposals.

Prerequisite skills and knowledge: Integration to find area between curves

Essential/useful library resources: none

Essential/useful computational resources: There is more flexibility in design if a computer graphing package and software for numerical or symbolic integration are available. However, these are not essential.

Example of an acceptable approach: A nice way to model the given design and generate other designs is to use curves which pass through the origin and through the point (1,1). Then the area between the curves will be exactly the fraction of the tile devoted to the inner color. The given design can be roughly modeled by $y = x^2$ and $y = \sqrt{x}$ on the interval [0,1]. This yields $\int_0^1 (\sqrt{x} - x^2)\,dx = \frac{1}{3}$. That is, 33% of the area is inside the curves.

One can achieve a pattern of intersecting circles by creating a design using the equations $(x-1)^2 + y^2 = 1$ and $x^2 + (y-1)^2 = 1$, again over the interval [0,1]. This yields

$$\int_0^1 \left[\sqrt{1-(x-1)^2} - (1 - \sqrt{1-x^2})\right] dx = \frac{\pi}{4} - (1 - \frac{\pi}{4}) = 0.57.$$

Hence, 57% of the total tile area is inside the curves.

Note: Students may use other means of finding areas. If you do not want this you may have to direct the students to using integration.

Questions for further exploration: A good open question related to this problem is to find a nonlinear design which will require the same amount of each color. There are many models to work from here. Students who find the project particularly interesting could also consider patterns created by alternating two tiles of different design (checkerboard style). Also, the two color limit could be lifted.

Special implementation suggestions: Students may need help getting started on this problem, either a discussion with the group at the outset as to how to approach the problem or the addition of some suggestions and/or hints as part of the problem statement. For example, students may not think to integrate to find area if not directed to do so. Thus, if you want to be certain the students use integration, you may have to specify so. Students may also need some help in thinking of using functions and their graphs to model the given design and to create new designs. Also, be prepared for a possible negative reaction or two. Most students have not had to do more than textbook-type problems, much less a project such as this one. A few may argue that this is not an appropriate assignment in a mathematics course. No doubt, there are instructors who feel the same way, but most who have used the project (both students and instructors) have enjoyed it immensely.

Special evaluation suggestions: The intent of this project is not only to have the students apply mathematics to the particular problem but also to present the designs they have created in a convincing way. Evaluation of the project should include consideration of the quality of the final presentation, and students should be aware of this from the outset.

Title: Minimizing the Area Between a Graph and Its Tangent Lines

Author: Steve Boyce, Berea College (problem suggested by R.C. Buck)

Problem Statement: Given a function f defined on [0, 1], for which of its non-vertical tangent lines T is the area between the graphs of f and T minimal? Develop an answer for three different nonlinear functions of your own choosing. Choose no more than one function from a particular class of functions (egs: polynomial, radical, rational, trigonometric, exponential, logarithmic). Carefully explain the reasoning leading to your conclusions. Looking back at your results, try to formulate and then verify any conjectures or generalizations they suggest. (Hint: stick to functions whose concavity doesn't change on [0,1].)

Information for the instructor only:

Problem abstract: Although the basic problem can be very simply stated, it invites several extensions involving various degrees of challenge. The examples students are asked to produce require multi-step solutions involving finding the equation of the tangent line at a general point, $(c, f(c))$, expressing the area A between the f and its tangent line as a function of c, and minimizing $A(c)$ on an appropriate interval. The examples students are most likely to choose will all result in $c = 1/2$ being the optimal location. This invites the conjecture that the tangent line at $c = 1/2$ always traps the minimum area. Possible extensions described below include the search for counterexamples and the verification that under certain conditions on f, $c = 1/2$ is the best location.

Prerequisite skills and knowledge : This project can be assigned any time after students have been introduced to the derivative applied to max/min analysis and the integral applied to finding the area between curves. For some of the possible extensions, it is necessary to understand the concavity/second derivative linkage and desirable to have access to software capable of numerical integration and root finding.

Essential/useful library resources: none

Essential/useful computational resources: none

Example of an acceptable approach: First, an example is presented using a function which seems typical of the kind students are likely to select. Following this example, several possible directions for project development are described and illustrated. Which, if any, of these directions is pursued depends on student initiative and insight and on choices the instructor makes regarding how much to include in the problem statement and what suggestions to offer once student work is underway.

Example 1: Suppose f defined by $f(x) = \sqrt{x}$ is selected as one of the three functions . The line tangent to f at $(c, \sqrt{c})$ has the equation $t_c(x) = \frac{1}{2\sqrt{c}}x + \frac{\sqrt{c}}{2}$, and the area between f and t_c , as a function of c, is given by $A(c) = \int_0^1 [t_c(x) - f(x)]dx = \frac{1}{4\sqrt{c}} + \frac{\sqrt{c}}{2} - \frac{2}{3}$. To finish, it is necessary to find where A achieves its minimum value on the interval (0,1]. Since $A'(c) = -\frac{1}{8\sqrt{c^3}} + \frac{1}{4\sqrt{c}} = \frac{1}{4\sqrt{c^3}}\left(-\frac{1}{2} + c\right) = 0$ only when $c = \frac{1}{2}$, this value of c provides the only critical point. It is evident from the First Derivative Test that $c = \frac{1}{2}$ is the absolute minimum on (0,1].

That $c = \frac{1}{2}$ locates the area-minimizing tangent line is not a peculiarity of this particular example. It is shown below using both analytic and geometric arguments that $c = \frac{1}{2}$ is always the best location provided the second derivative exists and is always positive or always negative on (0,1). The accessibility of these arguments makes it tempting to prompt students, if they require it, to conjecture the result and attempt its proofs. For some students, selecting three functions and clearly communicating reasoning which identifies the best tangent lines may be a good stopping point. For others, it may seem more appropriate to view the three examples as an invitation to generalization, conjecture and attempted verification.

Since the functions students seem most likely to choose have no inflection points on (0,1), the most natural conjecture is that $c = \frac{1}{2}$ is always the best location. (This conjecture would be less likely if [0,1] were replaced by [0,2π] in the problem statement!) For this reason, you may want to prompt them, via restatement of the problem or mid-project conference, to examine or search for a counterexample.

If students are to search for a counterexample, it is a good opportunity to encourage a graph sketching investigation, at least initially, as opposed to a focus on function formulas. The challenge is to show the conjecture false by sketching a graph, such as the one shown in Figure 1, for which $c = \frac{1}{2}$ is clearly not the optimum location.

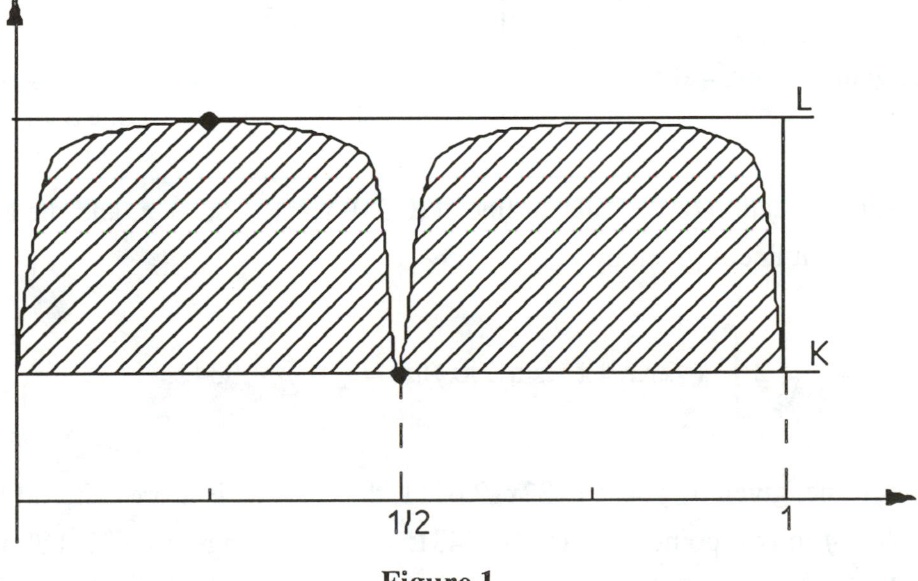

Figure 1

Students may not be able to find a function formula that provides a counterexample even if they have discovered or been shown a convincing graph. But the computations involved in verifying that a given function and tangent point do actually provide a counterexample can be sufficiently interesting that an instructor may wish to assign them as part of the project. Two such examples are given below, one of which does not assume familiarity with the calculus properties of trigonometric functions. In both examples, computer or calculator assistance is almost essential for finding intersection points of the function graph and various tangent lines, and it is at least convenient for evaluating some of the definite integrals that arise.

Example 2: The graphs in Figures 2 and 3 suggest that for the function $f(x)=\sin(2\pi x)$, the tangent line at $x = 0.4$ traps less area than does the tangent line at $x = 0.5$. If the instructor were to identify $\sin(2\pi x)$ as the source of a counterexample, some graph sketching might lead students to identify a value in the neighborhood of 0.4 or 0.6 as a target for an attempt at analytic verification.

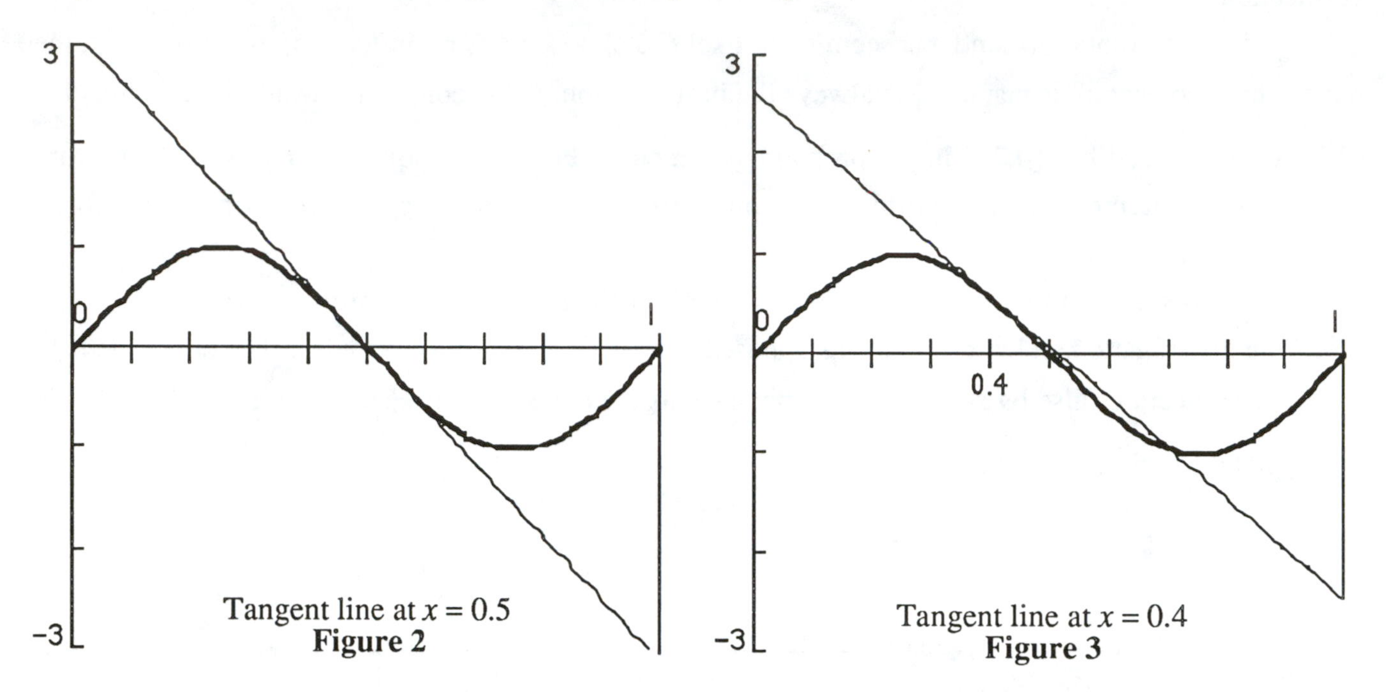

Tangent line at $x = 0.5$
Figure 2

Tangent line at $x = 0.4$
Figure 3

Since $f'(0.5) = -2\pi$, the equation of the tangent line at (0.5,0) is $y = -2\pi x + \pi$, and the area between the graphs of f and the tangent line is given by

$$2\int_0^{\frac{1}{2}} (-2\pi x + \pi - \sin(2\pi x))dx = \frac{\pi}{2} - \frac{2}{\pi} \approx .9342.$$

For the other tangent line, given by $y = -5.0832x+2.6211$, the analysis is more difficult because the line intersects the graph of f in two points, 0.4 and 0.70431. Since the tangent line is below the graph of f on (0.70431,1] and otherwise on or above it, the area between the two graphs is

$$\int_0^{0.70431} (-5.0832x + 2.6211 - \sin(2\pi x))dx + \int_{0.70431}^{1} (\sin(2\pi x) + 5.0832x - 2.6211)dx \approx 0.68266.$$

Example 3: This example is similar except that the function is a polynomial, $f(x) = -16x^3 + 24x^2 - 8x + 1$. Figures 4 and 5 show the graphs of f and two tangent lines, one located at $x = 0.5$ and the other at $x = 0.6$. The equation of the tangent line at (0.5,1) is $y = 4x - 1$, so the area between the two graphs can be seen to be

$$\int_0^{0.5} (-16x^3 + 24x^2 - 8x + 1 - 4x + 1)dx + \int_{0.5}^{1} (4x - 1 + 16x^3 - 24x^2 + 8x - 1)dx = 0.5.$$

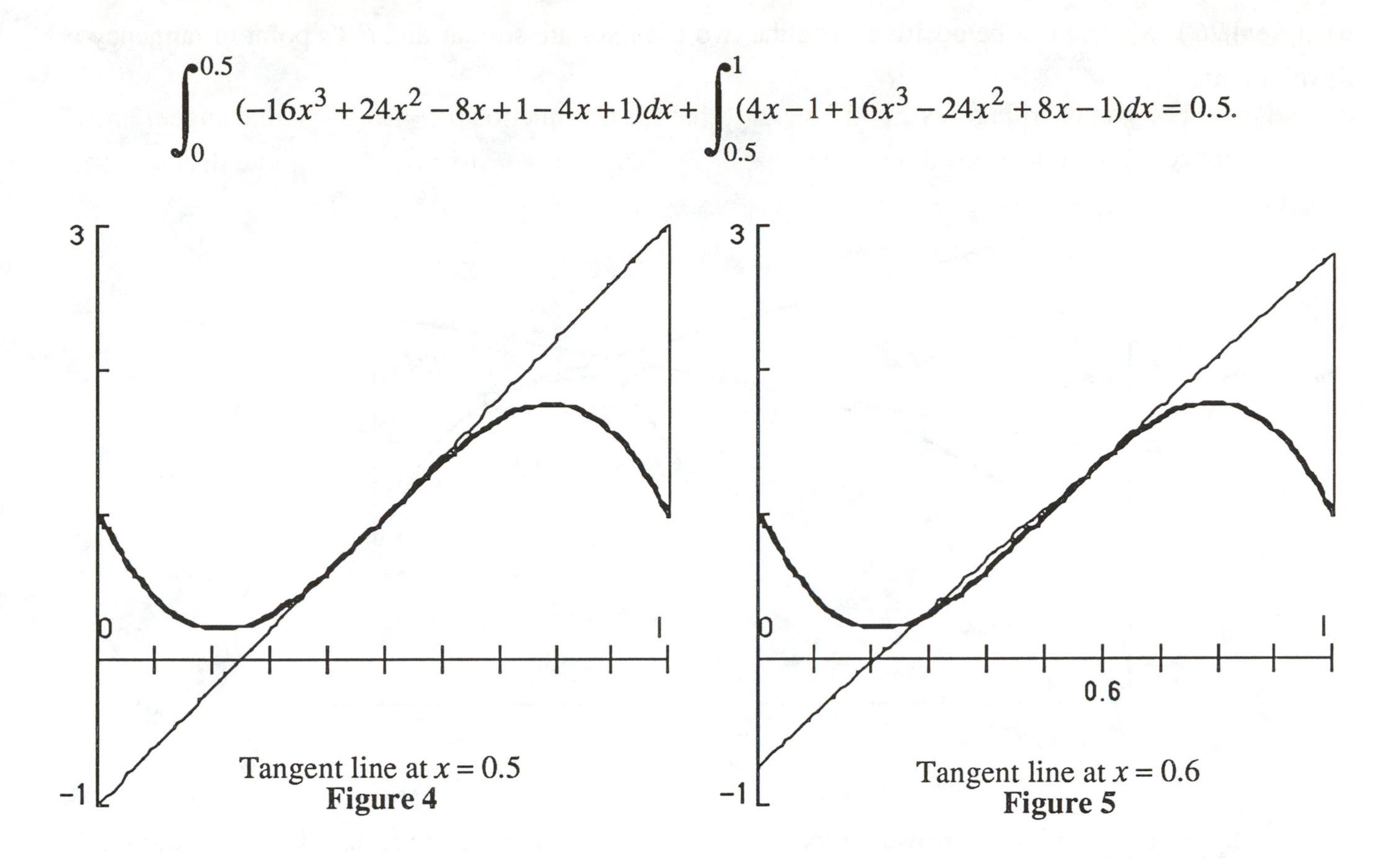

Tangent line at $x = 0.5$
Figure 4

Tangent line at $x = 0.6$
Figure 5

For the other tangent line, the equation is $y = 3.52x - 0.728$ and the points of intersection with f are located at $x = 0.6$ and $x = 0.3$. It follows that the area between f and the tangent line is

$$\int_0^{0.3} (-16x^3 + 24x^2 - 8x + 1 - 3.52x + 7.28)dx + \int_{0.3}^{1} (3.52x - .728 + 16x^3 - 24x^2 + 8x - 1)dx = 0.3992.$$

Even though 0.5 is not always the best location, if students' examples suggest to them that it "usually" is, then there may be some sparks of curiosity (waiting to be fanned) as to when it is guaranteed. There are at least two lines of reasoning, one geometric and the other analytic, that might lead students to discover the significance of the second derivative.

The geometric route is closely associated with efforts to sketch graphs for which 0.5 is not best. After seeing counterexamples like Figure 1, students might be asked: "Why can't you sketch a counterexample in which the graph of f is never above any of its tangent lines?" Something like the following argument might emerge: In Figure 6, consider the graph of a function f and two of its tangent lines, EC and FB where FB's point of tangency is $(0.5, f(0.5))$. Since all points of f are on or below both tangent lines, the region between f and either of the two lines can be divided into five subregions, $A1$ through $A5$ (see Figure 6). Let $A(f,EC)$ and $A(f,FB)$ represent the areas between f and the lines EC and FB respectively. It is clear from the figure that $A(f,EC) = A1 + A2 + A4 + A5$ and $A(f,FB) = A1 + A3 + A4 + A5$. It follows that $A(f,EC) - A(f,FB) = A2 - A3 =$ (the area of triangle AEF) – (the area of triangle *ABC*) which must be positive since the two triangles are similar and *FB*'s point of tangency is its midpoint.

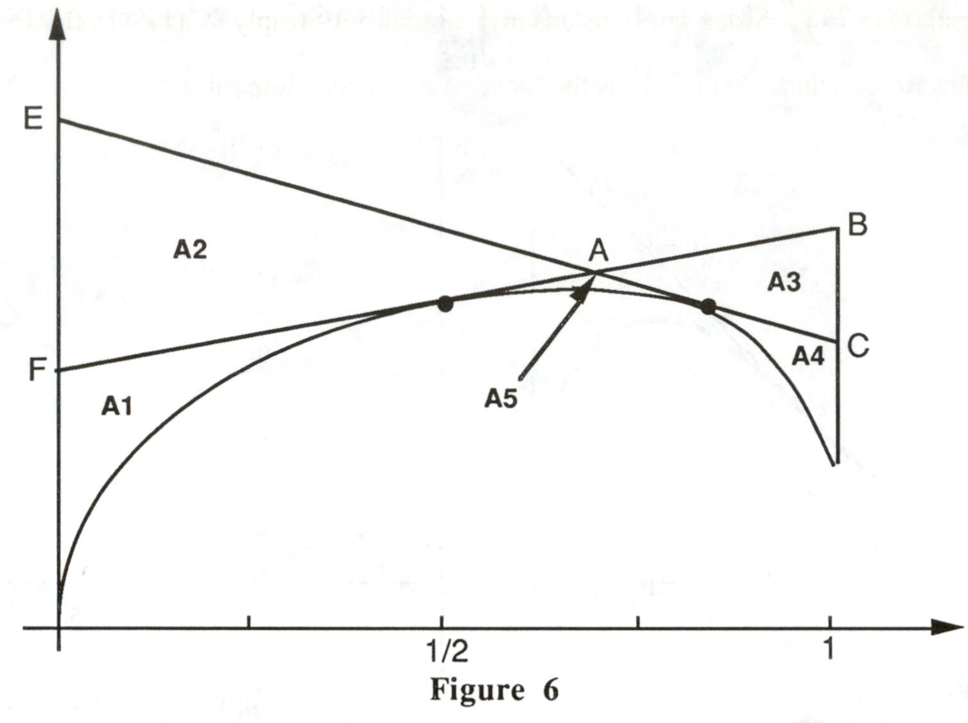

Figure 6

The analytic argument grows out of an effort to generalize the approach given in Example 1. After having worked through their three examples, students might be asked: "To what functions f can you apply the same basic method for showing that $c = 1/2$ is the best location? The area calculation is clearly easier if the graph of f on [0,1] is either entirely above or entirely below each tangent line (except for the point of tangency). So assume that about f for starters, and try to apply the same reasoning. What additional properties, if any, do you need to assume f has to make the method work?" Something like the following argument might emerge: The equation for the tangent line of f at $(c, f(c))$ is $y = f(c) + f'(c)(x-c)$. Since we can assume without loss of generality that the graph of f is never above its tangent line, the area between the two graphs, $A(c)$, is given by

$$A(c) = \int_0^1 (f(c) + f'(c)(x-c) - f(x))dx$$

$$= (f(c)x + f'(c)(\frac{x^2}{2} - cx))\Big|_0^1 - \int_0^1 f(x)dx$$

$$= f(c) + f'(c)(\frac{1}{2} - c) - \int_0^1 f(x)dx.$$

Assuming f'' exists, it follows that $A'(c) = f'(c) + f''(c)(\frac{1}{2} - c) + f'(c)(-1) = f''(c)(\frac{1}{2} - c) = 0$ only when

$c = \frac{1}{2}$ or $f''(c) = 0$. If we make the additional assumption that f'' is never zero on (0,1), then the only interior critical point is $c = \frac{1}{2}$. Since previous assumptions about f imply $f''(x) < 0$, the First Derivative Test can be applied to conclude that $c = \frac{1}{2}$ is the location of the absolute minimum.

One difficulty students may perceive in generalizing their approach to the examples is that without knowing f, they have no hope of knowing how to perform the integration $\int_0^1 f(x)\,dx$ which arises in the expression for $A(c)$. As a way of raising the same issue in a less abstract setting, it may be useful to assign an example in which the antiderivative of f is not an elementary function.

Example 4: Suppose $f(x) = e^{x^2}$. Then the equation of the line tangent at $(c, f(c))$ is

$$y = 2c\, e^{c^2} x + (1 - 2c^2)\, e^{c^2}.$$

Since f is never below any of its tangent lines,

$$= \int_0^1 [f(x) - t_c(x)]dx = \left[2ce^{c^2} \frac{x^2}{2} + (1 - 2c^2)e^{c^2} x \right] \Big|_0^1 - \int_0^1 e^{x^2} dx$$

$$= (c + 1 - 2c^2)e^{c^2} - \int_0^1 e^{x^2} dx.$$

At this point students may realize that since the remaining integral is simply a constant and the next step is to calculate the derivative of A, it is never necessary to calculate the integral's value — whatever it is, its derivative is zero. To finish, note that

$$A'(c) = (1 - 4c)\, e^{c^2} + (c + 1 - 2c^2)\, 2c\, e^{c^2} = (1 - 2c)(1 + 2c^2)\, e^{c^2} = 0 \text{ only when } c = \frac{1}{2},$$

and apply the First Derivative Test .

Conjectures we expect that some students will make: It would be nice to see students conjecture that $c = \frac{1}{2}$ will always minimize the area and then make some attempt at determining the conditions necessary to prove their conjecture.

Questions for further exploration: Pursuing a full or even partial solution to the $c = \frac{1}{2}$ conjecture is a good exercise for strong students. A discussion of this is included in the section "Example of an acceptable approach."

Title: Riemann Sums, Integrals, and Average Values

Authors: Eugene Herman and Charles Jones, Grinnell College

Problem Statement: The goal of this project is for you to develop and explain the use of Riemann sums in application problems. The focus of your writing should be on clear descriptions and justifications of your methods. There will be three groups of questions for you to consider.

Question Group 1 - Average Temperature

When we say, "The average temperature today was 60 degrees," we clearly intend the single number 60 to represent the entire range of temperatures for the day. It is not so clear, however, how this number is to be computed. If we have a finite sample of temperatures, we can simply compute their average. For example, from the table of hourly temperatures (see Figure 1) in Des Moines, Iowa, on June 10, 1990, we can compute the average temperature by adding the 24 numbers and dividing by 24.

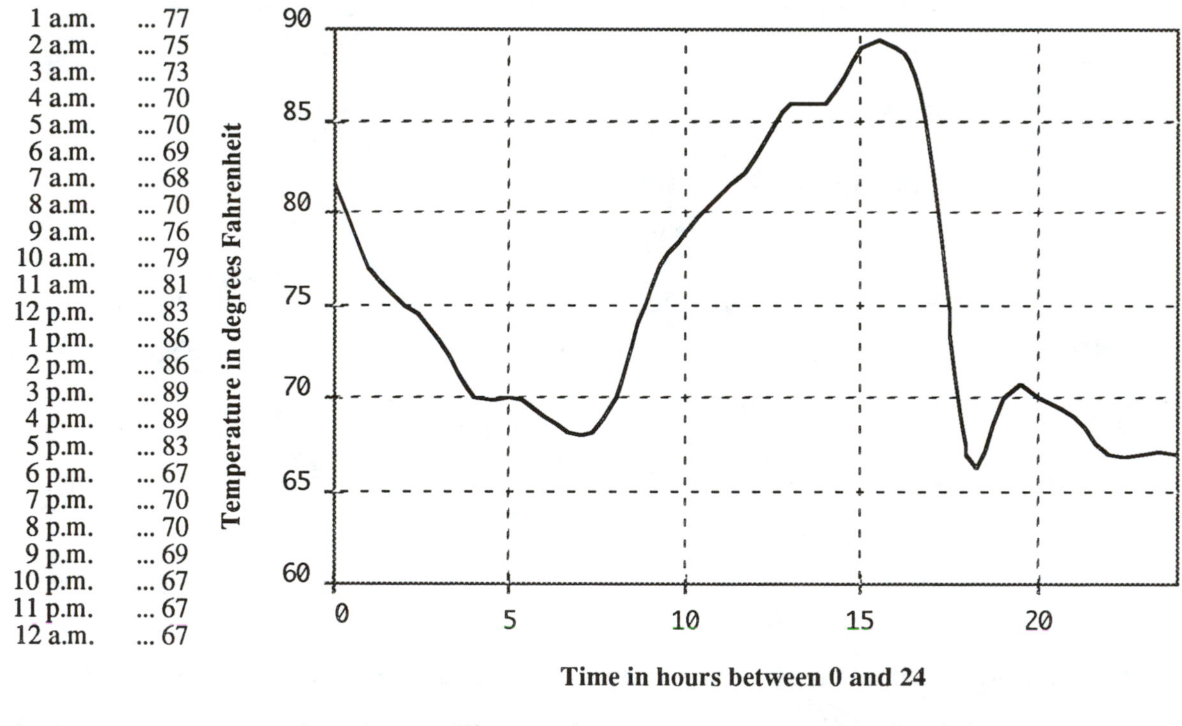

Time	Temperature
1 a.m.	... 77
2 a.m.	... 75
3 a.m.	... 73
4 a.m.	... 70
5 a.m.	... 70
6 a.m.	... 69
7 a.m.	... 68
8 a.m.	... 70
9 a.m.	... 76
10 a.m.	... 79
11 a.m.	... 81
12 p.m.	... 83
1 p.m.	... 86
2 p.m.	... 86
3 p.m.	... 89
4 p.m.	... 89
5 p.m.	... 83
6 p.m.	... 67
7 p.m.	... 70
8 p.m.	... 70
9 p.m.	... 69
10 p.m.	... 67
11 p.m.	... 67
12 a.m.	... 67

Figure 1

Question 1a) Compute this average. How would you compute the average temperature if the temperatures were recorded every half-hour instead of every hour? Explain why this average will usually not be the same as the average of the hourly temperatures.

Our intuition suggests that measuring the temperature more often should lead to a better estimate of the average temperature. So let's take this idea to the limit. Suppose the temperature at time t in hours ($0 \le t \le 24$) is $T = f(t)$ in degrees Fahrenheit. (See the graph in Figure 1.) If the temperature is measured n times in 24 hours, say at times $t_1, t_2, \cdots, t_n$, the average of these temperatures is

$$\frac{1}{n}\sum_{i=1}^{n} f(t_i).$$

Suppose the times t_i are equally spaced so that $t_i - t_{i-1} = \Delta t_i = \frac{24}{n}$. Then

$$\frac{1}{n}\sum_{i=1}^{n} f(t_i) = \frac{1}{24}\sum_{i=1}^{n} f(t_i)\Delta t_i \qquad (1)$$

which has the form of a Riemann sum multiplied by 1/24.

Question 1b) The graph in Figure 1 represents the temperature function f whose values at each hour are exactly the temperatures in the table. Use the graph to compute the Riemann sum of f with $n = 6$ and f evaluated at right endpoints of subintervals. Then multiply by 1/24. (The answer should be close to, but not equal to, your answer in 1a.) If you had used $n = 24$ instead of $n = 6$, you would have gotten exactly the answer in 1a; explain why.

Question 1c). Give a careful argument to explain why

$$\frac{1}{24}\int_0^{24} f(t)\,dt$$

should be the true average temperature over the 24-hour time period. Use Riemann sums and limits in your argument.

This group of questions addressed a special case of the more general concept of the average of a function, which is used in the remaining two groups of questions. The *average* of a function f on the interval $[a, b]$ is

$$\frac{1}{b-a}\int_a^b f(x)\,dx. \qquad (2)$$

Question Group 2 - Velocity and Distance

Question 2a) Approximate the average velocity and the distance traveled of an object whose velocity function is described by the *table* in Figure 2.

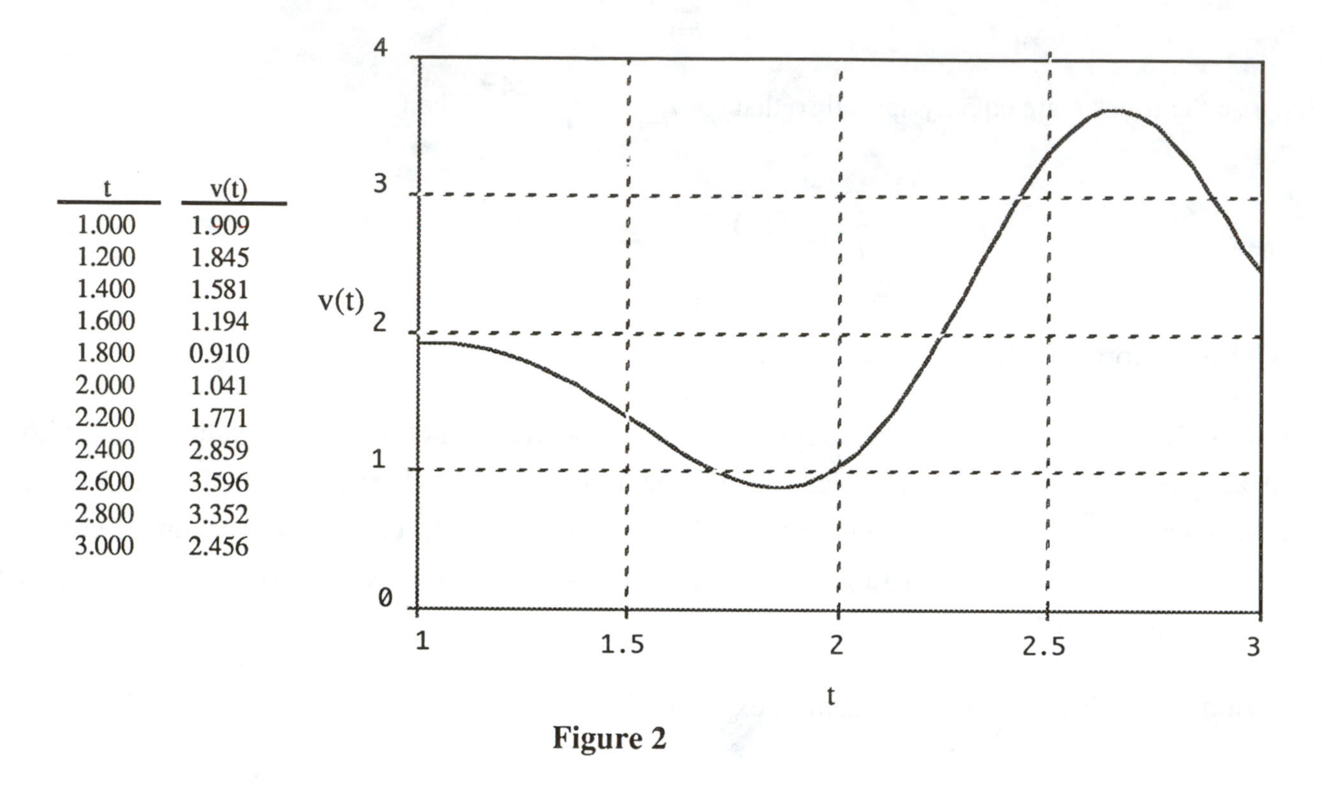

t	v(t)
1.000	1.909
1.200	1.845
1.400	1.581
1.600	1.194
1.800	0.910
2.000	1.041
2.200	1.771
2.400	2.859
2.600	3.596
2.800	3.352
3.000	2.456

Figure 2

Question 2b) Suppose you are given the velocity $v(t)$ of an object at all times t, where $a \leq t \leq b$. Assume $v(t) \geq 0$ for all t. Use Riemann sums and limits, as in question group 1, to derive a formula for the distance traveled. Explain and justify your derivation. Also describe how the concepts of distance and area are related. (This should follow from your derivation.)

Question 2c) The average velocity, according to formula (2), is

$$\frac{1}{b-a}\int_a^b v(t)\,dt.$$

Use the formula you derived in question 2b to explain why average velocity is also equal to $\Delta s/\Delta t$ (change in position divided by change in time).

Question Group 3 - Mass and Center of Mass

We consider a straight rod of length L. We will assume that the density ρ of the rod varies only along its length, so we will measure ρ in units of the form mass/length (e.g., grams/cm). We denote position on the rod by x (with $0 \le x \le L$), so $\rho = \rho(x)$ is defined on the interval $[0, L]$.

Question 3a) Approximate the mass of the rod whose density is described by the *table* in Figure 3.

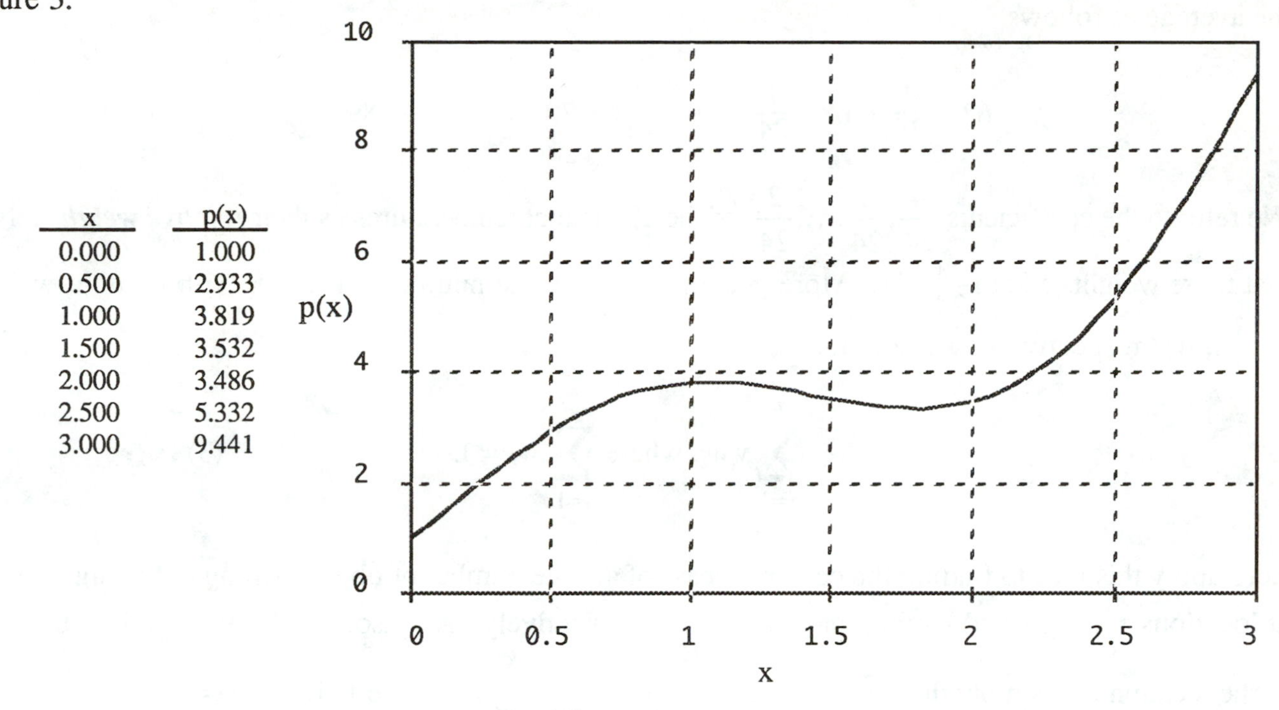

x	p(x)
0.000	1.000
0.500	2.933
1.000	3.819
1.500	3.532
2.000	3.486
2.500	5.332
3.000	9.441

Figure 3

Question 3b) Use Riemann sums and limits to explain why the mass, M, of the rod is given by

$$M = \int_0^L \rho(x)\, dx.$$

Question 3c) From question 3b, M = area under the graph of ρ between $x = 0$ and $x = L$. On the graph of ρ in Figure 3, draw the rectangles whose area corresponds to the Riemann sum you computed in question 3a.

Question 3d) A rod made of a homogeneous material, that is, one for which ρ is a constant function, has a particularly easy to find center of mass: the midpoint of the rod. We can also use the notion of average value to compute the center of mass; we simply find the average x coordinate. Compute this average,

$$\frac{1}{L}\int_0^L x\, dx,$$

and verify that it is the midpoint.

To see how to find the center of mass when the rod is made of a nonhomogeneous material, we return to our initial example of temperatures. Notice that when you averaged the 24 numbers in the table of Figure 1, there were only 13 distinct numbers. So we can group equal numbers together to compute the average as follows:

$$67 \cdot \frac{4}{24} + 68 \cdot \frac{1}{24} + 69 \cdot \frac{2}{24} + 70 \cdot \frac{5}{24} + \ldots + 89 \cdot \frac{2}{24}\ .$$

We refer to the coefficients $\frac{4}{24}, \frac{1}{24}, \ldots, \frac{2}{24}$ of the 13 distinct temperatures as their *relative weights*. Note that these weights add to $\frac{24}{24} = 1$. More generally, to average numbers $y_1, \ldots, y_k$ with relative weights $w_1, \ldots, w_k$, respectively, we compute

$$\sum_{i=1}^{k} y_i w_i, \text{ where } \sum_{i=1}^{k} w_i = 1. \tag{3}$$

Let's apply this idea to finding the center of mass of a finite number of objects strung out along an x axis at locations $x_1, \ldots, x_k$ and having mass $m_1, \ldots, m_k$, respectively. A reasonable relative weight to assign to the location x_i is simply the relative mass $\frac{m_i}{m}$ where $m = \sum_{i=1}^{k} m_i$. Note that $\sum_{i=1}^{k} \frac{m_i}{m} = 1$. The center of mass is thus

$$\bar{x} = \sum_{i=1}^{k} x_i \frac{m_i}{m}. \tag{4}$$

Question 3e) Approximate the center of mass of the rod whose density is described by the table in Figure 3.

Question 3f) Use Riemann sums and limits to derive a formula for the center of mass of a rod made of a nonhomogeneous material. Explain and justify your derivation. (As in question 3b, you are given the density function ρ.)

Information for the instructor only:

Problem abstract: This project is designed to give students practice in deriving integral formulas for applications by taking a limit of Riemann sums. The particular problems addressed were chosen to be accessible to all students, not just those with some specific background in science. Although the problems may appear to be easy, students do not find them so. In textbooks, they are rarely asked to produce and defend their own derivations of integral formulas. Each group of questions starts with a concrete computational question that should help them begin thinking about Riemann sums.

Prerequisite skills and knowledge: The definite integral as a limit of Riemann sums; at least one application, such as a volume formula found by taking a limit of approximating Riemann sums.

Essential/useful library resources: none

Essential/useful computational resources: any calculator

Example of an acceptable approach: The most crucial questions are 1c, 2b, 3b, and 3f. In all of these, we hope to see a reasonably complete Riemann sum argument of the type they have seen in the textbook and in class. For example, an acceptable answer to 2b might be the following.

Divide $[a, b]$ into n equal subintervals with endpoints $a = t_0 < t_1 < ... < t_n = b$. So

$$t_i - t_{i-1} = \Delta t = \frac{b-a}{n}.$$

If Δt is small, v will not change much on each subinterval. So the distance traveled during such a short time period will be approximately $v(t_i)\Delta t$ (distance = rate · time). So the total distance will be approximately

$$\sum_{i=1}^{n} v(t_i)\Delta t.$$

By the definition of the integral,

$$\lim_{n\to\infty} \sum_{i=1}^{n} v(t_i)\Delta t = \int_a^b v(t)dt.$$

Also, the approximations

$$\sum_{i=1}^{n} v(t_i)\Delta t$$

get closer to the exact distance as $n \to \infty$. So, the exact distance equals

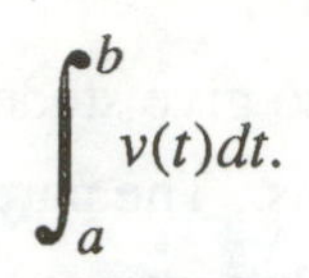

An acceptable answer to question 2c might simply be

$$\int_a^b v(t)dt \text{ is the distance traveled, } \Delta s.$$

The change in time, Δt, is $b - a$. So, $\dfrac{1}{b-a}\displaystyle\int_a^b v(t)dt = \dfrac{\Delta s}{\Delta t}$.

Questions for further exploration:

1. In question 2c you were asked to explain why

$$\frac{1}{b-a}\int_a^b v(t)dt = \frac{\Delta s}{\Delta t}.$$

   Use the Fundamental Theorem of Calculus to explain why this equality is true.

2. The desired formula in answer to question 3f is

$$\frac{1}{M}\int_0^L x\rho(x)dx, \text{ where } M = \int_0^L \rho(x)dx$$

   is the mass in question 3b. Rewrite this formula for center of mass in the form

$$\frac{1}{L}\int_0^L x\, w(x)dx, \text{ where } \frac{1}{L} = \int_0^L w(x)dx = 1. \qquad (5)$$

That is, find the function w. Express $w(x)$ in terms of $\rho(x)$ and the average value of ρ. Use the discussion of weighted averages in question 3 and the formula in question 3d to give a plausibility argument for (5).

References/bibliography/related topics: All the usual integral formulas for application.

Special evaluation suggestions: On the exam covering this material, give a question that requires a limit of Riemann sum argument to derive an integral formula to solve a concrete problem which cannot be solved by any integral formula the students already know.

Title: The Ice Cream Cone Problem

Author: Matt Richey, St. Olaf College

Problem Statement: You are to place a sphere of ice cream into a cone of height 1.

(1) What radius of the sphere will give the most volume of ice cream inside the cone (as opposed to above the cone) for a cone with a base angle of 30°?

(2) What percent of this sphere of most volume lies inside the cone?

You will need to be thoughtful in choosing the variables which will best help you answer this question. When you have determined the optimal radius, be sure to make an accurate drawing of your sphere in the cone to insure the reasonableness of your result.

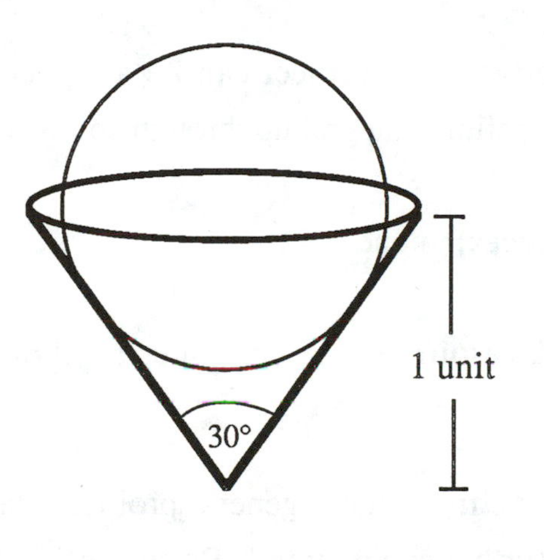

Figure 1

Information for the instructor only:

Problem abstract: This is essentially a complicated max/min problem, and a rather silly application of techniques that are very important in solving more realistic problems in which one needs to optimize some geometric relationship between unlike objects. Students will find this project difficult because they will have to determine from sketches expressions involving more than one variable, and then compute with those expressions. In order to make progress, they will have to choose to let some of the variables act as parameters, at least temporarily. In the process of completing this project, they will develop their skills with problems involving geometric relationships, algebraic and trigonometric calculations, optimization, tangent lines, and volumes of solids of revolution. They will have to pose and solve a chain of smaller problems, employing a variety of known calculus techniques, to obtain solutions to the stated questions. One other important aspect of this project is that the students could choose approaches that are technically correct, but if followed, make it impossible to answer the question at hand. They will need to be warned about this, and it will be important for them to confer with you regularly so that you can provide appropriate guidance.

Prerequisite skills and knowledge: This project can be assigned to students any time after they have studied the derivative and the definite integral up through solids of revolution.

Essential/useful library resources: none

Essential/useful computational resources: A computer algebra system with a graphics package would be useful.

Example of an acceptable approach: (A more general problem statement is suggested below as (1) under the heading "Questions for further exploration." Responses for the alternative problem statement are provided here. These alternative statement responses have been placed in braces {...} along with the solution to the original problem statement.)

The intended meaning of "volume inside the cone" is the volume of the sphere from its bottom up to a plane level with the top of the cone. To derive the formula for this volume the students should define the desired portion of the sphere as a solid of revolution and use the standard textbook approach. (This is a good place for the students to be more general in approach than they usually are: at the beginning it is not clear how they are going to need to use this formula, so they should try to find a general formula now and specialize it later.) If they let the sphere have radius "r" and denote the distance from the bottom of the sphere to the plane as "b," they then need the volume of the solid defined by $x = \sqrt{r^2 - (y-r)^2}$ revolved about the y-axis as y goes from 0 to b. (See Figure 2.)

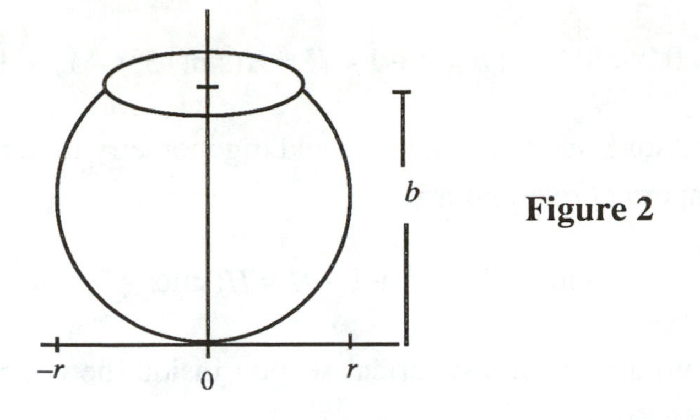

Figure 2

From this it follows that $V(b) = \int_{y=0}^{b} \pi(\sqrt{r^2-(y-r)^2})^2\,dy = \pi\left[rb^2 - \frac{1}{3}b^3\right]$. Deriving this formula for the volume of a partial sphere, with b as the independent variable and r as a parameter, is important because some of the later computations can become quite complicated without it. (You might choose to give this formula to the students.) Now, the cone could be described by the rotation of the line $y = kx$ about the y-axis, but it is more useful for this problem to describe it using the base angle of the cone, 30°. {In attacking the alternative problem call the base angle 2α, with $0 \le \alpha \le \frac{\pi}{2}$.}

We can accurately picture the resulting object in two dimensions, as shown in Figure 3, with "H" as the height of the center of the sphere from the bottom of the cone. The two diagrams in Figure 3 illustrate that the center of the sphere might be above or below the top of the cone. (The students may need help to see that the most reasonable interpretation of the sphere "sitting inside the cone" is not that it is sitting on the lip of the cone but that the sphere is touching the side of the cone tangentially, as shown in Figure 3.) Intuitively, one of the two parameters r and H is free with the value of the second being determined by the other. Also, the value of the variable b will need to be described in terms of r and H. {Intuitively, two of the three parameters r, α, and H are free with the value of the third being determined by the other two. Also, the value of the variable b will need to be described in terms of r, α, and H.}

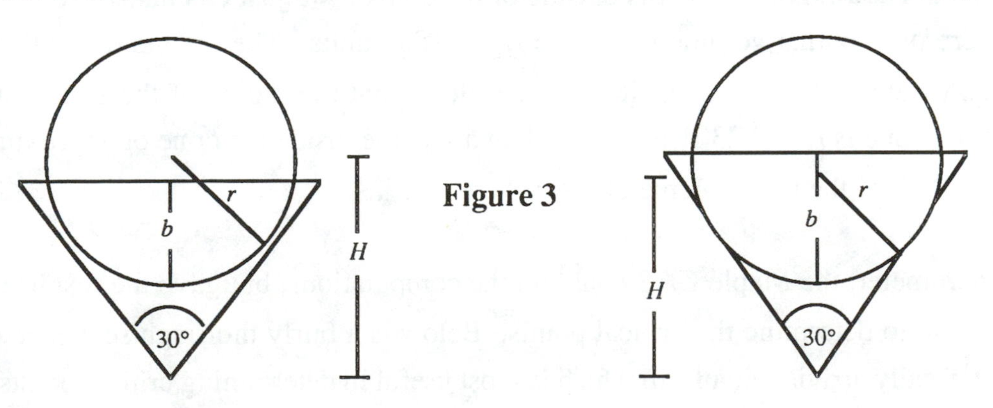

Figure 3

Students may believe at first that the expression for b depends on whether the center of the sphere is above or below the top of the cone. They will find that this is not so; the expression for b is the same in either case. With a base angle of 30°, the important relations are:

$$r = H\sin(15°) \approx 0.2588H \text{ and } b = r + 1 - H = H(\sin(15°) - 1) + 1 \approx 1 - 0.7412H.$$

(You may need to encourage students to use sketches and trigonometry to derive these relations.) {With an arbitrary α, the important relations are

$$r = H\sin\alpha \text{ and } b = r + 1 - H = H(\sin\alpha - 1) + 1.\}$$

Using the formula for the volume of the spherical section inside the cone, eliminating b and r, they should get

$$V(H) \approx \pi\left[(0.2588H)(1 - 0.7412H)^2 - \tfrac{1}{3}(1 - 0.7412H)^3\right] = \pi\left[0.2779H^3 - 0.933H^2 + H - \tfrac{1}{3}\right].$$

$$\left\{V(H) = \pi\left[(H\sin\alpha)(H(\sin\alpha - 1) + 1)^2 - \tfrac{1}{3}(H(\sin\alpha - 1) + 1)^3\right].\right\}$$

The students should now determine the possible values of H. Intuitively, $b \le 2r$, since once the sphere is completely inside the cone its radius and volume will only get smaller. From this it follows that $H \ge \frac{1}{1 + \sin 15°} \approx 0.7944 \left\{H \ge \frac{1}{1 + \sin\alpha}\right\}$, so H has a lower bound. In addition, the condition that the sphere be tangent to the inside of the cone imposes an upper bound on H. If we define "s" as the slant height of the cone then it follows that $H = \frac{s}{\cos(15°)} = \frac{1}{\cos^2(15°)} \approx 1.0718 \left\{H = \frac{s}{\cos\alpha} = \frac{1}{\cos^2\alpha}\right\}$ is the last value at which the cone and the sphere are tangent. Hence, $0.7944 \le H \le 1.0718$ $\left\{\frac{1}{1 + \sin\alpha} \le H \le \frac{1}{\cos^2\alpha}\right\}$ gives the domain of the inside volume function. (Students may also need your guidance in determining these bounds.) Knowledge of the domain is important because there are two critical points of the volume function and only one of them is in the domain interval.

It is quite easy to calculate $V'(H) \approx \pi[0.8337H^2 - 1.866H + 1]$, yielding the critical points: $H_c \approx 1.3492$ or 0.8890. The first of these falls outside of the domain we just calculated, so the critical point giving the sphere of maximal volume should be $H_c = 0.889$ units. The second derivative test quickly verifies this, as would a plot of $V(H)$. It quickly follows that the radius of the sphere with maximal volume inside the cone is $r_c \approx 0.2301$ units; yielding a volume inside the cone of approximately 0.0426 cubic units, or 83.4% of the total volume of that sphere.

{With α as a parameter, the Maple CAS could do the computations, but gave the results in a form that made it impossible to determine the critical points. Below is a fairly thorough sequence of equivalent forms of $V'(H)$, finally arriving at a form which is most useful in determining critical points.

$$V'(H) = \pi[\sin\alpha(H(\sin\alpha - 1) + 1)^2 + 2H\sin\alpha(H(\sin\alpha - 1) + 1)(\sin\alpha - 1) - (H(\sin\alpha - 1) + 1)^2(\sin\alpha - 1)]$$

$$= \pi\,[2H\sin\alpha(H(\sin\alpha - 1) + 1)(\sin\alpha - 1) + (H(\sin\alpha - 1) + 1)^2]$$
$$= \pi\,[2H^2\sin\alpha(\sin\alpha - 1)^2 + 2H\sin\alpha(\sin\alpha - 1) + H^2(\sin\alpha - 1)^2 + 2H(\sin\alpha - 1) + 1]$$
$$= \pi[H(\sin\alpha - 1)[H(2\sin^2\alpha - \sin\alpha - 1) + 2(\sin\alpha + 1)] + 1]$$
$$= \pi[H(\sin\alpha - 1)(H(2\sin^2\alpha - \sin\alpha - 1)) + 2H\sin^2\alpha - H\sin\alpha - H + H\sin\alpha - H + 1]$$
$$= \pi[H(\sin\alpha - 1)(H(2\sin^2\alpha - \sin\alpha - 1)) + H(2\sin^2\alpha - \sin\alpha - 1) + H(\sin\alpha - 1) + 1]$$
$$= \pi[(H(\sin\alpha - 1) + 1)(H(2\sin^2\alpha - \sin\alpha - 1) + 1)].$$

Setting $V'(H) = 0$ yields two critical points: $H = \dfrac{1}{1 - \sin\alpha}$ and $H = \dfrac{1}{1 + \sin\alpha - 2\sin^2\alpha}$. For the first candidate, since $1 - \sin\alpha \leq 1 - \sin^2\alpha = \cos^2\alpha$, it follows that this value of H is outside the domain of V since $H \geq \dfrac{1}{\cos^2\alpha}$. For the second critical point, $H \geq \dfrac{1}{1 + \sin\alpha}$, since $1 + \sin\alpha - 2\sin^2\alpha \leq 1 + \sin\alpha$. Also, $H \leq \dfrac{1}{\cos^2\alpha}$, since $\cos^2\alpha \leq \cos^2\alpha + \sin\alpha = 1 + \sin\alpha - 2\sin^2\alpha$. Hence the desired critical point is $H_c = \dfrac{1}{1 + \sin\alpha - 2\sin^2\alpha}$. The typical first or second derivative tests do not work very nicely in determining whether H_c is a maximum or a minimum. However, if students observe that $V(H)$ is a cubic polynomial in H and remember the shape of a cubic with a positive third-degree term, it follows that H_c, being the lesser of the two critical values, yields a local maximum value of $V(H)$. They will need to verify that the coefficient of H^3 is positive, but this can be done pretty quickly:

$$V(H) = \pi\left[(H\sin\alpha)(H(\sin\alpha - 1) + 1)^2 - \tfrac{1}{3}(H(\sin\alpha - 1) + 1)^3\right]$$
$$= \pi\left[H^3(\sin\alpha(\sin\alpha - 1)^2 - \tfrac{1}{3}(\sin\alpha - 1)^3) + H^2(2\sin\alpha(\sin\alpha - 1) - (\sin\alpha - 1)^2) + H - \tfrac{1}{3}\right].$$

So, for H_c to give the desired maximum value of $V(H)$, it is only necessary that $\sin\alpha(\sin\alpha - 1)^2 - \frac{1}{3}(\sin\alpha - 1)^3 = (\sin\alpha - 1)^2(\frac{2}{3}\sin\alpha + \frac{1}{3})$ be positive. This condition holds because $\sin\alpha \geq 0$ for $0 < \alpha \leq \frac{\pi}{2}$. So the dimension of the sphere which gives the most volume of ice cream inside the cone for a cone with fixed base angle 2α is $r_c = H_c \sin\alpha = \dfrac{\sin\alpha}{1 + \sin\alpha - 2\sin^2\alpha}$. And that completes the alternate problem!}

Conjectures we expect that some students will make: There is more to this problem than ice cream.

Questions for further exploration:

(1) A more general and more difficult problem statement is: You are to place a sphere of ice

cream into a cone of height 1. What radius of the sphere will give the most volume of ice cream inside the cone for a cone with a fixed base angle, 2α? The solution to this one is given along with the stated problem solution.

(2) If you allow the base angle to vary, what is the closest that the center of the sphere of most volume will get to the bottom of the cone?

Here we need to find a minimum value of H_c with α as the variable. Having found that $H_c = \dfrac{1}{1 + \sin\alpha - 2\sin^2\alpha}$, it follows that $\dfrac{dH_c(\alpha)}{d\alpha} = \dfrac{4\sin\alpha\cos\alpha - \cos\alpha}{(1 + \sin\alpha - 2\sin^2\alpha)^2}$. Setting this to 0, we get $\cos\alpha(4\sin\alpha - 1) = 0$. So H_c has critical points at $\cos\alpha = 0$ and $\sin\alpha = \frac{1}{4}$. If $\cos\alpha = 0$, then $a = \frac{\pi}{2}$ and H_c and $\dfrac{dH_c(\alpha)}{d\alpha}$ are not defined. If $\sin\alpha = \frac{1}{4}$, then $H_c = \dfrac{1}{1 + \sin\alpha - 2\sin^2\alpha} = \dfrac{1}{1 + \frac{1}{4} - 2[\frac{1}{4}]^2} = \frac{8}{9}$.

To verify that this is the minimum of H_c, the students could use either the first or second derivative test.

This result is a pleasant surprise; there is no reason to expect such a "nice" rational number. Notice also that the radius of the lowest sphere of maximal volume is $r = H\sin\alpha = [\frac{8}{9}][\frac{1}{4}] = \frac{2}{9}$. Hence the bottom of the lowest sphere of maximal volume is $H - r = \frac{8}{9} - \frac{2}{9} = \frac{2}{3}$ units above the vertex of the cone.

(3) What is the limiting behavior of H_c as α approaches 0 and $\frac{\pi}{2}$?

As α gets close to 0, H_c approaches 1 and you get a very small sphere almost completely enclosed within the cone. As α gets close to $\frac{\pi}{2}$, H_c gets very large and you get a very large sphere (the radius gets arbitrarily large) almost completely outside of the cone.

(4) If you allow the base angle to vary, is the sphere with maximal volume inside the cone ever centered exactly at a height of 1?

It quickly follows that $H = 1$, $b = r = \sin\alpha$, and $V'(H) = \pi[(\sin^2\alpha)(2\sin\alpha - 1)]$. Hence, if $\alpha = 30°$ ($2\alpha = 60°$), the sphere of maximal inside volume has its center at a height of 1. In this case 50% of the sphere is inside the cone.

(5) This problem could be done in two dimensions: you could place a parabola inside an angle. (If you place a circle inside an angle, you run into the interesting problem that there is no simple closed form expression for the area of the circle inside the triangle.)

References/bibliography/related topics: This project is clearly related to others that seem more rooted in applied mathematics: designing a phonograph needle to get the best contact with a record groove; designing a ball point pen; others where we need to optimize some relationship between different-shaped objects. A student interested in engineering would profit from finding and reading information about such applications.

Special implementation suggestions: The instructor or a teaching assistant will likely need to give considerable attention to students working on this project. There are many places where students could have trouble with the algebra or trigonometry required for this project, especially in its alternative form. It will require a group of students with perseverance to complete this successfully. There are a few places where it would be appropriate for you as the instructor to require that students verify with you the correctness and appropriate form of intermediate results. Otherwise, they could spend too much time fruitlessly spinning their computational wheels. I would suggest that you at least have them confirm their results on the formulas $V(H)$ and $V'(H)$ before moving on. You may need to give them some guidance to insure they make reasonable progress. I have made parenthetical comments where I thought guidance to be most appropriate. As I noted in the description of the solution, a computer algebra system with graphics capability might prove useful at times, but students need to realize that most computer algebra systems do computations in forms that are often not most useful to us: they should not depend on a CAS to enlighten them, only to help them check out ideas and calculations.

MULTIVARIATE CALCULUS

The following projects are ones whose prerequisites include topics usually covered in a multivariate calculus course. All of the projects heavily use functions of several variables, partial derivatives, and/or multiple integration. The first two projects involve applications of multivariate extrema theory to various settings. *Waste Container Construction* applies the theory to a specific "real world" problem while *Own Your Own Function of Two Variables* uses extrema theory to analyze functions of two variables. The *Three Cylinder Intersection Problem* is primarily an exercise in visualizing a surprisingly difficult and intriguing 3-dimensional solid. Both *Own Your Own Function of Two Variables* and *Three Cylinder Intersection Problem* require students to evaluate double integrals. Finally, students can be exposed to the classic "steepest ascent" optimization algorithm in *Gradient Method Optimization.*

Projects from previous sections that are also appropriate in a multivariate calculus course are *Designing a Pipeline With Minimum Cost, Crankshaft Design,* and *Valve Cover Design. Designing a Pipeline With Minimum Cost* becomes a multivariate optimization problem if the two elbow joint restriction is removed, while the other two projects can be handled nicely using parametric equations.

MULTIVARIATE CALCULUS

Title: Waste Container Construction

Author: John Ramsay, College of Wooster

Problem Statement: Locate a waste container (trash dumpster) in your area, study its construction and determine its volume. Given the general shape, volume and method of construction of the inspected container, determine the dimensions such a container should have in order to minimize the cost of construction (material and labor). You should maintain the volume and basic shape of the original container.

The following facts will help you in calculating costs:

- The sides, back and front are to be made from 12 gage (0.1046 in. thick) steel sheets which cost 66¢/sq. ft.
- The base is to be made from a 10 gage (0.1345 in. thick) steel sheet which costs 84¢/sq. ft.
- Lids are ordered separately and cost approximately $40.00 regardless of dimensions.
- Cutting the steel sheets to the correct size is done with a shearing machine which works like a giant meat cleaver. Thus, the length of the cut does not change the overall cost of construction. The same is true for bending the sheets.
- Welding costs approximately 15¢/ft. for material and labor combined.

The final report should include:

- a description of the entire process
- the location and a sketch of the model waste container
- *all* details of construction—ignoring any detail in the model must be justified
- justification of any assumptions made to simplify and commentary on how those assumptions may or may not affect the final result
- a recommendation as to whether or not it is worth changing the current dimensions.

Show all your work in solving the problem and explain your conclusions completely.

Information for the instructor only:

Problem abstract: This project is designed to give students the opportunity to complete a problem from its beginnings. It will force students to consider the conflict between modeling a real problem closely and maintaining a mathematical model which is solvable. The need to make construction assumptions and modeling simplifications is an important experience that most calculus students do not get. In addition it gives them the opportunity to see the mathematics they have studied applied to a "real world" problem.

Prerequisite skills and knowledge: Multivariate theory of extrema.

Essential/useful library resources: none

Essential/useful computational resources: Under most methods of construction the solution to the model will involve solving a nonlinear system of equations and hence it will be necessary that students have access to equipment suited for solving such equations (symbolically, graphically or numerically).

Example(s) of an acceptable approach and/or actual student work: Included here are solutions to two standard container types. There are many other shapes and sizes manufactured. More complicated shapes create much more complicated solutions and also require further assumptions on fixed dimensions. The second example gives some sense of these complications and assumptions. In the examples presented, all parts other than the sides and base of the container are ignored assuming that these parts will not change significantly (if at all) without relatively large changes in the overall dimensions. Students should be expected to list all such extraneous points and comment on why they have been included or not.

Example 1:

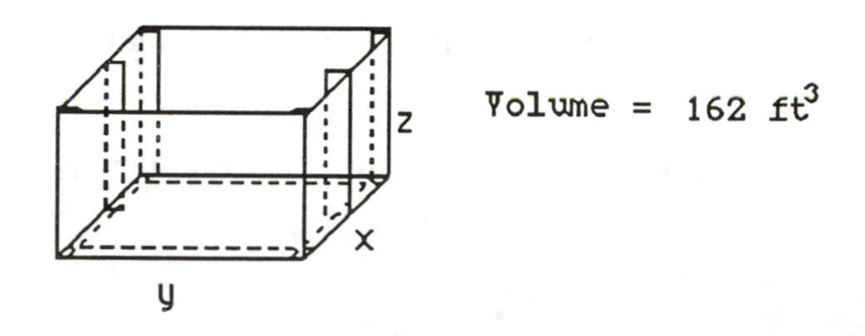

The front and back are each a single sheet while each side is made up of two sheets. These two sheets are equal in size and are attached down a seam in the middle of the side of the container. All vertical seams require an extra 1 inch overlap from the side sheet and horizontal seams (attaching the base) use 1 inch from the upright sheet.

area of base sheet: xy

area of front sheet: $y(z + \frac{1}{12})$

area of back sheet: $y(z + \frac{1}{12})$

area of half of side: $(\frac{x}{2} + \frac{2}{12})(z + \frac{1}{12})$

We obtain the following expression for cost:

$$\text{cost} = 0.84xy + 0.66[2y(z + \tfrac{1}{12}) + 4(\tfrac{x}{2} + \tfrac{2}{12})(z + \tfrac{1}{12})] + 0.15(2y + 2x + 6z) + \text{fixed cost}$$

subject to the constraint: $xyz = 162.$

We obtain the solution $x = 6.5$ ft. (78 in.), $y = 6.5$ ft. (78 in.), $z = 3.84$ ft. (46 in.) with variable cost = \$111.79. This is not far from the 72 in. by 72 in. by 54 in. dimensions of one of the models currently manufactured and the cost savings is quite modest as the modeled variable cost of the current container is \$112.51.

It is important to note that containers which involve slanted tops or front require further assumptions on the problem. Angles formed and/or altitudes of triangular sheets of steel will need to be fixed in order to keep the problem from becoming too complicated to solve. The next example illustrates this:

Example 2:

The following sketch gives the dimensions of a slanted front waste container.

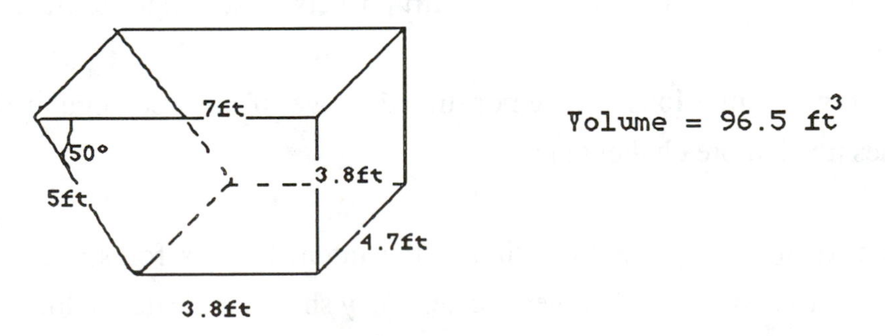

Note: In this example we do not have the overlap at the seams.

We will assume the angle at the top, front is to remain fixed at 50° due to handling requirements. Then we label the container as follows:

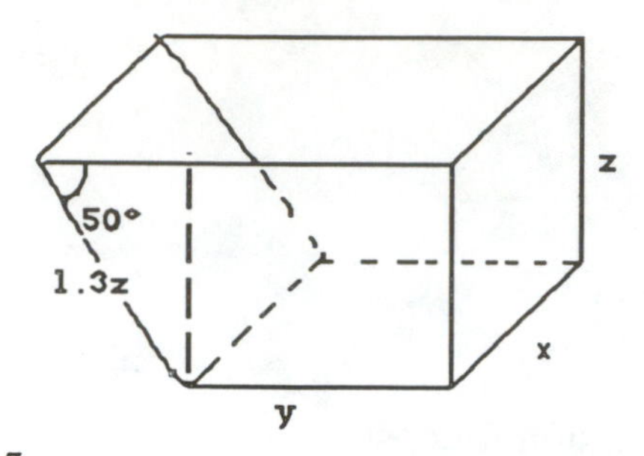

where $1.3z$ comes from $\sin(50°) = \frac{z}{?}$.

We obtain

$$\text{cost} = 0.84xy + 0.66\left[xz + 2(yz + \frac{1}{2}0.83z \cdot z) + 1.3z \cdot x\right] + 0.15[2y + 2x + 2z + 2(1.3z)]$$

subject to the constraint: $xyz + \frac{1}{2}(0.83z)zx = 96.5$.

Combining by eliminating y yields

$$\text{cost} = 1.1694xz + \frac{81.06}{z} + \frac{127.38}{x} + 0.5655z + \frac{28.95}{xz} + 0.3x.$$

This function has a minimum of \$73.96 at $x = 5.6$ ft. (67 in.), $y = 3.5$ ft. (42 in.), $z = 3.5$ ft. (42 in.). The given container had a cost of \$74.50, hence very minimal savings would be achieved.

Conjectures we expect that some students will make: Students will likely include other parts of the container in their model. (For example, lids, lift pockets, base supports, etc...)

Questions for further exploration: As demonstrated above, if non-rectangular sheets are involved the problem becomes much more challenging.

Special implementation suggestions: Firstly, the intent here is for students to determine the general shape from a waste container that they locate. They should indicate the location of the container so that verification of the accuracy of their model can be made, if necessary. Secondly, finding the solution of the system of partial derivatives may require substantial effort and care even with the help of a technology tool. Students should be warned to check each other's computations at this stage to eliminate errors. Finally, it needs to be clear to the students that they are to consider all the detail of the container and its construction.

Special evaluation suggestions: One of the intentions of a problem such as this one is to have the student analyze and justify the assumptions they have made in constructing their particular model. The written or oral presentation of their solution should include those assumptions. One possibility is to have students approach the problem as if they had been hired as a consulting firm. They then submit a written report of their findings in language that a person without knowledge of calculus could understand as well as a second document detailing the mathematics.

Title: Own Your Own Function of Two Variables

Authors: Eugene Herman and Anita Solow, Grinnell College

Problem Statement: Select one of the four functions given by the expressions below (or use one selected by your instructor), and investigate its behavior thoroughly by all the means at your disposal. In particular, find its

(a) absolute maximum and minimum values;
(b) local maximum and minimum points and saddle points;
(c) double integral;
(d) average value.

Do as much as you can by hand and as much as you can by computer. Compare results that you get both ways to confirm that they agree. Where you cannot find exact answers, approximate. Check that your numerical results are reasonable. For example, use a graph to confirm visually the local and absolute extrema, saddle points, and average value.

In your report, you may omit routine details of the computations, but you must include all of the following: responses to parts (a)-(d) and explanations of how you found them; at least one graph and a discussion of how it supports your answers; comparisons of results obtained by computer with results obtained by hand.

1. $f(x, y) = 2x^2y + xy^2 - 6xy$, $-1 \le x \le 3$ and $-1 \le y \le 3$.

2. $f(x, y) = 3\sqrt{x + y + 4} - 2\sqrt{x} - \sqrt{y}$, $0 \le x \le 6$ and $0 \le y \le 6$.

3. $f(x, y) = \sin x + \sin y + \sin(x + y)$, $0 \le x \le 2\pi$ and $0 \le y \le 2\pi$.

4. $f(x, y) = xe^{-(x^2 + y^2)/8}$, $-6 \le x \le 6$ and $-6 \le y \le 6$.

Information for the instructor only:

Problem abstract: This project gives students an opportunity to employ all the skills and understanding they have developed in their study of functions of two variables. It is a very flexible project in that they need only go as far in their hand computations as they can, and they require only the most basic features of a graphing package for functions of two variables. For example, if the algebra required to find the critical points is too hard for them, their graph is likely to guide them to these points; if the software will only compute approximations to a double integral, that is good enough. A crucial part of the project is the requirement that they check that their results are reasonable and that they compare results obtained by more than one technique. This is a good test of students' depth of understanding.

Prerequisite skills and knowledge: Absolute and local maxima and minima for functions of two variables on a rectangle; double integrals over a rectangle; experience in using a graphing package for functions of two variables.

Essential/useful library resources: none

Essential/useful computational resources: Any graphing package for functions of two variables.

Example of an acceptable approach: For all four functions, all four questions can be answered completely using only hand techniques. However, many students will have difficulty finding the critical points or the double integral. Here is an abbreviated version of a student report for function 3.

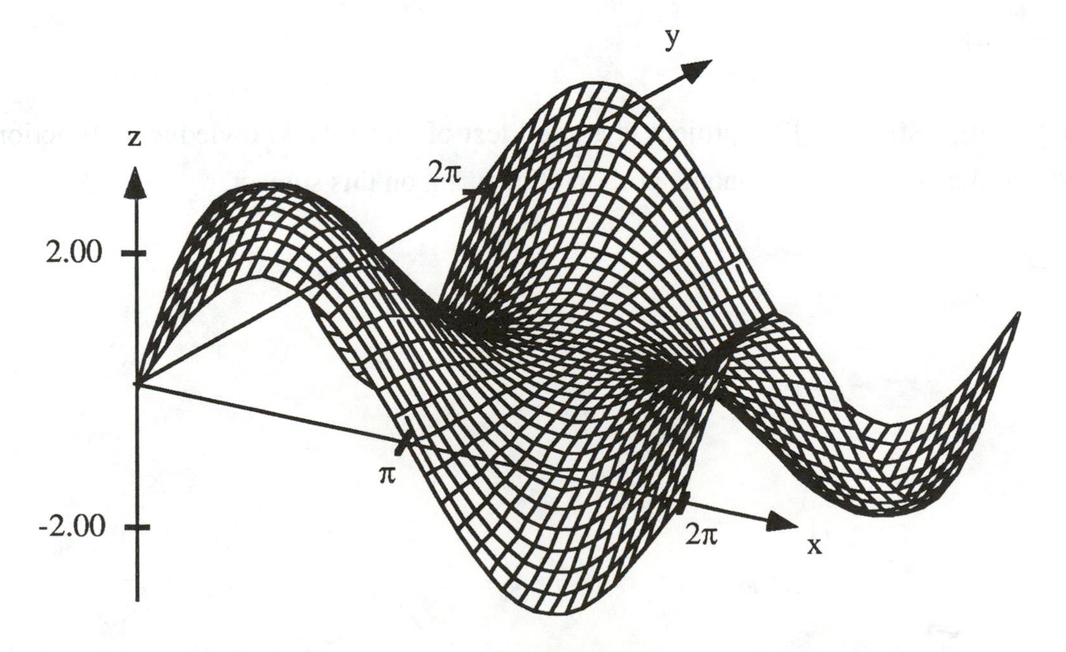

3. $f(x, y) = \sin x + \sin y + \sin(x + y)$

Find the critical points:

$$\frac{df}{dx} = \cos x + \cos(x + y) = 0$$

$$\frac{df}{dy} = \cos y + \cos(x + y) = 0.$$

Either note that $\cos x = \cos y$ (and hence $x = y$ or $x = 2\pi - y$), or use the picture or a numerical routine to guess that the critical points are at $\left(\frac{\pi}{3}, \frac{\pi}{3}\right)$, (π,π), and $\left(\frac{5\pi}{3}, \frac{5\pi}{3}\right)$ and confirm by plugging into the critical point equations. The discriminant at (π,π) is zero, although (π,π) looks like a saddle point in the picture. Depending on the definition you use, (π,π) is either indeterminate or a saddle point (since every neighborhood of (π,π) contains points (x, y) where $f(x, y) > f(\pi,\pi)$ and points (x, y) where $f(x, y) < f(\pi,\pi)$). At $\left(\frac{\pi}{3}, \frac{\pi}{3}\right)$, f has a local and absolute maximum; at $\left(\frac{5\pi}{3}, \frac{5\pi}{3}\right)$, f has a local and absolute minimum. The absolute maximum and minimum values are $\frac{\pm 3\sqrt{3}}{2}$. (Along the boundary, $|f(x, y)| \leq 2$.) The double integral and the average value of f are zero. The symmetries in the expression for f and in the graph of f suggest this conclusion.

Questions for further exploration: Build a model of the graph of the function out of any materials you like (clay, paper, string, straws, etc.). Use this model to demonstrate and confirm your conclusions, such as the absolute and local extrema. Mark the key features on your model.

Special implementation suggestions: Change the project from year to year by replacing the functions with new ones.

Special evaluation suggestions: This project is a good test of students' knowledge of functions of two variables and could be used as a take-home portion of an exam on this subject.

Title: Three Cylinder Intersection Problem

Author: John Ramsay, College of Wooster

Problem Statement:

a) Investigate the solid enclosed by the three cylinders $x^2 + y^2 = 1$, $x^2 + z^2 = 1$ and $y^2 + z^2 = 1$. Carefully sketch *and* describe the solid and compute its volume.

b) What happens to the solid described in a) if the radii of the cylinders are varied? Before considering the general case $x^2 + y^2 = a^2$, $x^2 + z^2 = b^2$, and $y^2 + z^2 = c^2$; you may want to investigate some specific cases such as $a = b = c = 2$ and/or $b = c = 1$, $a > 1$ or $a < 1$. Again you should describe the solids in question in whatever way seems appropriate and at least set up the integrals which give their volume.

Information for the instructor only:

Problem abstract: This problem, though simply stated, is quite a challenge in three dimensional visualization. It is intended as a means of pushing students to improve their ability to visualize a complex solid and communicate their "picture" to others. It is particularly well suited for encouraging group interaction. The volume calculation is included primarily to assist in forcing a clear understanding of the solid.

Prerequisite skills and knowledge: An introduction to quadric surfaces and multiple integration with cylindrical coordinates.

Essential/useful library resources: none

Essential/useful computational resources: The entire problem can certainly be worked by hand, although some of the integration is quite complicated. A symbolic or numerical integration package (for multiple integrals) might help but the primary importance is in set-up, not computation. A good 3-d graphics package is also helpful but not at all necessary.

Example of an acceptable approach: Describing the simplest case, $a = b = c = 1$ is by no means trivial. The shape is somewhat spherical; in fact, traces along $x = 0$, $y = 0$ and $z = 0$ are unit circles. However, the solid bulges out to a "vertex" at the eight points $(\pm\frac{1}{\sqrt{2}}, \pm\frac{1}{\sqrt{2}}, \pm\frac{1}{\sqrt{2}})$ forming seams in the planes $y = \pm x$, $z = \pm x$ and $z = \pm y$. The traces in these planes are not circles as they pass through the above points as well as the units on the coordinate axes. A sketch of the $y = x$ trace is given here:

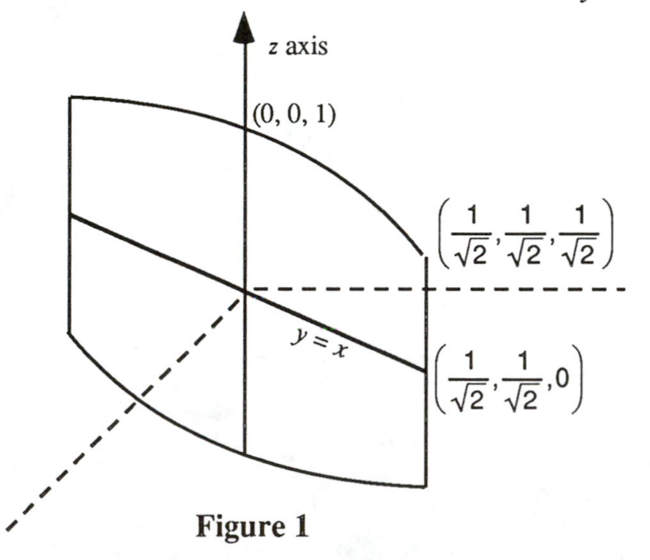

Figure 1

There are 12 four sided (curved sides) concave faces formed, each spanning one of the equators. A first octant view of the solid is given in Figure 2.

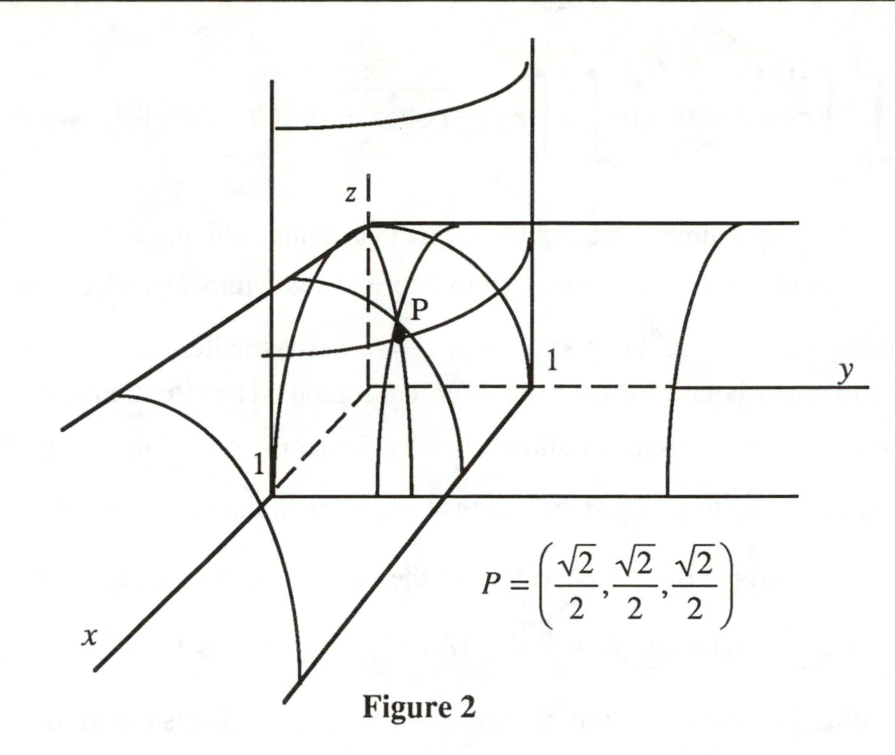

Figure 2

A useful perspective, especially in setting up the volume integrals, is a birds-eye view. Viewing from atop the z axis on the intersection of the cylinders $x^2 + z^2 = 1$ and $y^2 + z^2 = 1$ yields Figure 3 below. The solid formed by the intersection of these two projects onto the square $[-1,1] \times [-1,1]$. The shading lines in the figure indicate which cylinder is forming the top of the solid in question. If we now add in the third cylinder, it simply reduces the square projection to a circular projection as seen in Figure 4 below.

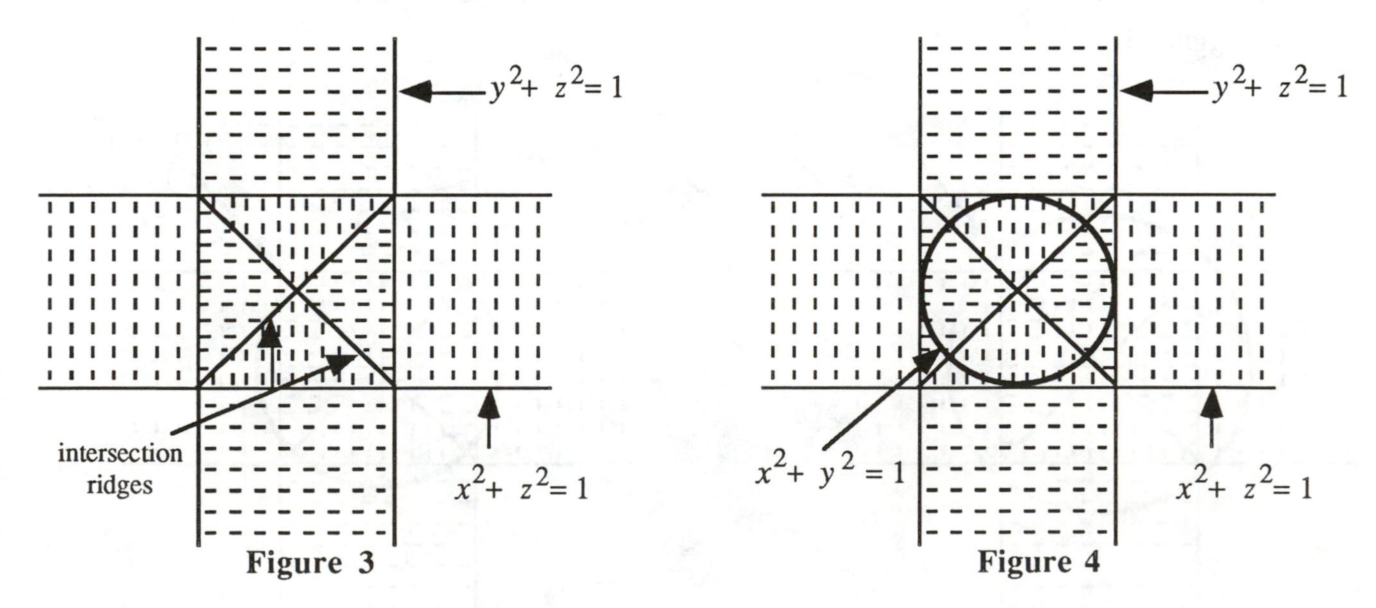

Figure 3 **Figure 4**

The volume is obtained from the integral

$$16\int_0^{\frac{\pi}{4}}\int_0^1\int_0^{\sqrt{1-x^2}} r\,dz\,d\,r d\,\theta = 16\int_0^{\frac{\pi}{4}}\int_0^1 r\sqrt{1-r^2\cos^2\theta}\,dr\,d\,\theta = 16-8\sqrt{2} \approx 4.6863.$$

(This result confirms that the volume is are slightly larger than a unit sphere.)

In the volume calculation, we let θ range from 0 to $\frac{\pi}{4}$ using symmetry. We have $x^2 + z^2 = 1$ and $y^2 + z^2 = 1$ intersecting along $y = x$. For $y < x$, $x^2 + z^2 = 1$ has a smaller z coordinate than does $y^2 + z^2 = 1$ and hence is the upper bound for the solid over that region. The situation is reversed for θ from $\frac{\pi}{4}$ to $\frac{\pi}{2}$ but in this particular case symmetry allows us to perform only one of the integrals.

Another interesting way to describe the solid is to begin with the cube with vertices $(\pm\frac{1}{\sqrt{2}}, \pm\frac{1}{\sqrt{2}}, \pm\frac{1}{\sqrt{2}})$. Note that all the edges of this cube are on the surface of the desired solid. For example: $(x, \frac{1}{\sqrt{2}}, \frac{1}{\sqrt{2}})$, $-\frac{1}{\sqrt{2}} \le x \le \frac{1}{\sqrt{2}}$ satisfies $x^2 + z^2 \le 1$, $y^2 + z^2 = 1$, $x^2 + y^2 \le 1$ and is thus on the boundary of the solid. The solid is then obtained from the cube by rounding the faces out to the units on the axes. This will cause an inward cusping along the edges of the cube while through the vertices the trace will be as described in Figure 1 above. Tracing at the center of each face is in one of the coordinate planes and hence must be a unit circle.

The more general cases simply modify the work already performed in the first part. A few comments on these is all that will be included here as there are so many directions that could be pursued.

If one of the cylinders has a radius significantly larger than the other two, this cylinder will have no impact on the solid whatsoever as it will completely contain the solid described by the two smaller ones (see Figure 5).

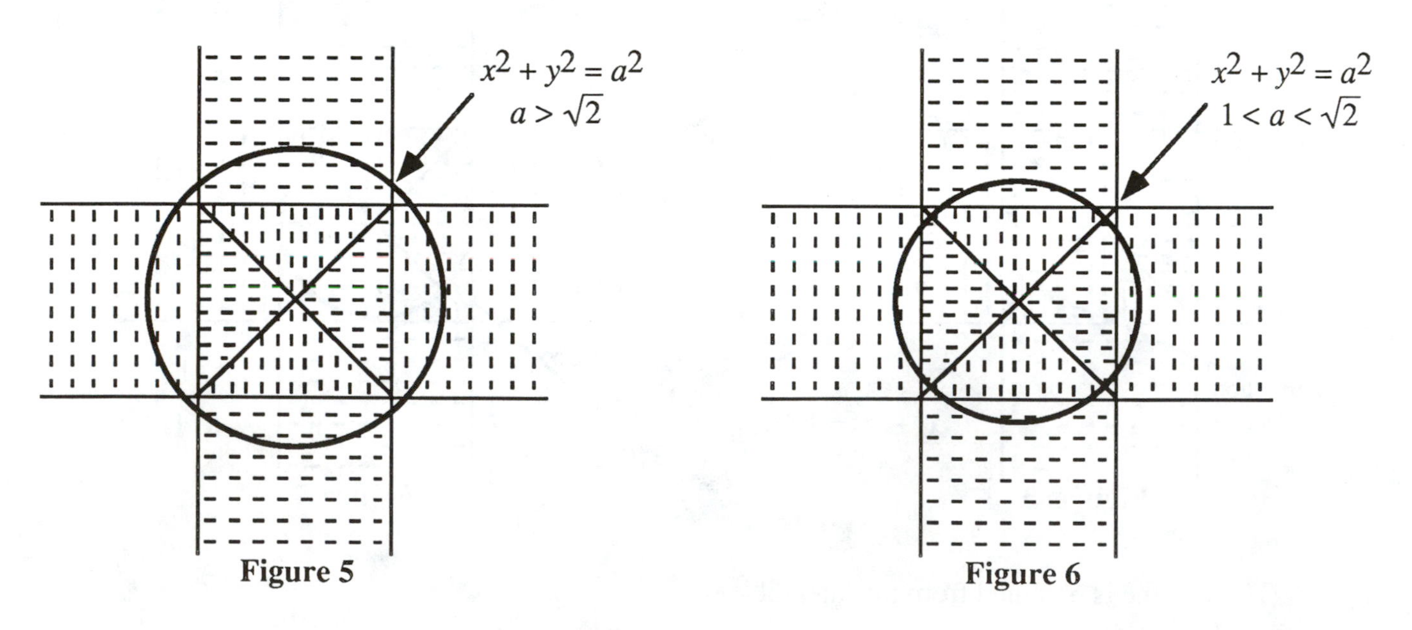

Figure 5 **Figure 6**

Begin with the case $a = b = c = 1$ and allow a to increase while b and c remain equal to 1. The change in the solid is simply in a stretching in the size of faces described by the b and c cylinders and a shrinking in the size of the faces described by the a-cylinder. It is easiest to visualize this by looking at Figure 3 and imagining the circle increasing in size (see Figure 6). When the radius, $a = \sqrt{2}$, the a-cylinder will just touch the b and c-cylinder intersection point lying on the xy-plane. At this point, the a-cylinder has become too large to play any role in the enclosed solid, as was noted in Figure 5.

The volume computation in the $b = c = 1$, $1 < a \leq \sqrt{2}$ case is complicated only slightly. As the a-cylinder increases in size, the region above $\theta = 0$ to $\frac{\pi}{4}$ is still bounded by $x^2 + z^2 = 1$ but the bound of the region in the xy-plane is no longer $r = 1$. As a increases, the initial boundary (beginning at $\theta = 0$ will be the straight line from (1,0,0) to (1,1,0) and, depending on a, will then move to the circle $x^2 + y^2 = a^2$. (A sketch of this planar region is given below.)

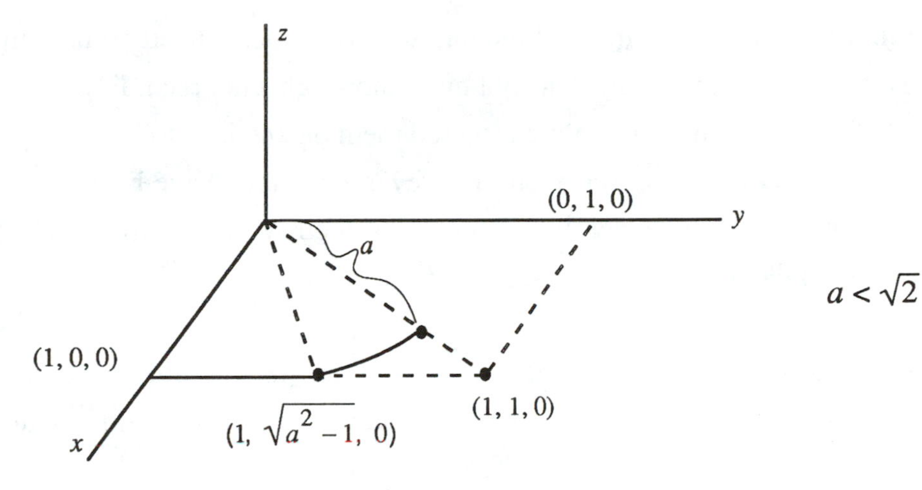

Figure 7

The volume can be computed as follows:

$$V = 16\int_0^1\int_0^{\sqrt{a^2-1}}\int_0^{\sqrt{1-x^2}} dz\,dy\,dx + 16\int_{\cos^{-1}\left(\frac{1}{a}\right)}^{\frac{\pi}{4}}\int_0^{a}\int_0^{\sqrt{1-x^2}} r\ dzdrd\theta.$$

Please see discussion of the general case below with regards to the evaluation of these integrals and further discussion of the general case.

Conjectures we expect that some students will make: Many students will find the special case $a = b = c = 1$ sufficiently difficult and may stop there. Some may consider only cases $a = b = c$. Some may assume symmetry in each octant (which is correct) but not realize that two separate integrals are required to compute the volume in each octant. (One with the z integral 0 to $\sqrt{1 - x^2}$, the other z integral 0 to $\sqrt{1 - y^2}$.)

Questions for further exploration: Classifying the problem for all a, b and c is quite a chore, though accessible to strong students who turn on to the problem. The integrals on the more general cases are also quite interesting. For example, in the volume integrals given above for the case $b = c = 1$, $1 \leq a \leq \sqrt{2}$, the first integral can be done by hand but the second requires a numerical approximation technique. Students could use a numerical integration package to study the integral for various values of a, noting that when $a = 1$, $V = 4.6863$ (done in part a)) and when $a \geq \sqrt{2}$, volume is simply determined by the intersection of two cylinders. A similar problem arises for $b = c = 1$, $a < 1$ with volume given by

$$16\int_0^{\frac{\pi}{4}}\int_0^a r\sqrt{1-r^2\cos^2\theta}\,dr\,d\theta.$$

Special implementation suggestions: Students will very likely need some help getting going on this one. They should be encouraged to obtain a thorough understanding of the solid and some suggestions at the beginning may be required to prod them on and also to help them see the complexity involved in providing a complete description. However, once they have begun to get the picture, they should be able to progress independently. A check of their work in part a) is highly recommended before they move on to the more general cases in part b).

Special evaluation suggestions: Students' ability to visualize three dimensional objects is greatly varied and this problem relies heavily on this skill. Evaluation should be done with the understanding that even the simplest case will be quite challenging for many students.

A sketch could be required but it is not an easy thing to produce. Try it yourself!

Title: Gradient Method Optimization

Author: John Ramsay, College of Wooster

Problem Statement: Techniques and algorithms for optimizing linear functions of several variables are widespread and very effective. On the other hand, algorithms for optimizing nonlinear problems are much harder to come by and those that exist are not nearly as effective as their linear counterparts. The general approach for nonlinear optimization problems is to follow an approximation algorithm similar to Newton's method for finding zeros of functions of one variable. Recall that Newton's method gives us a means of successively stepping from one approximation of the desired zero to a (hopefully) better approximation. A single formula gives both the direction to move (left or right) and the distance (step size) to move. We need to make similar decisions in creating an algorithm for solving nonlinear optimization problems involving more than one independent variable. Given an approximation to the extremum, we must determine how to choose a better approximation by choosing a direction and a step size. For functions of several variables, direction means a vector and step size, representing a distance, is a positive real number. We want an algorithm of the form $\boldsymbol{x}_{i+1} = \boldsymbol{x}_i + \lambda_i \boldsymbol{p}_i$ where $\boldsymbol{x}_i$ is the old approximation, $\boldsymbol{x}_{i+1}$ is the new approximation, λ_i represents step size and $\boldsymbol{p}_i$ is a unit vector in the desired direction. In the formula, $\boldsymbol{x}_i$, $\boldsymbol{x}_{i+1}$ and $\boldsymbol{p}_i$ are vectors and λ_i is a positive real number (see Figure 1).

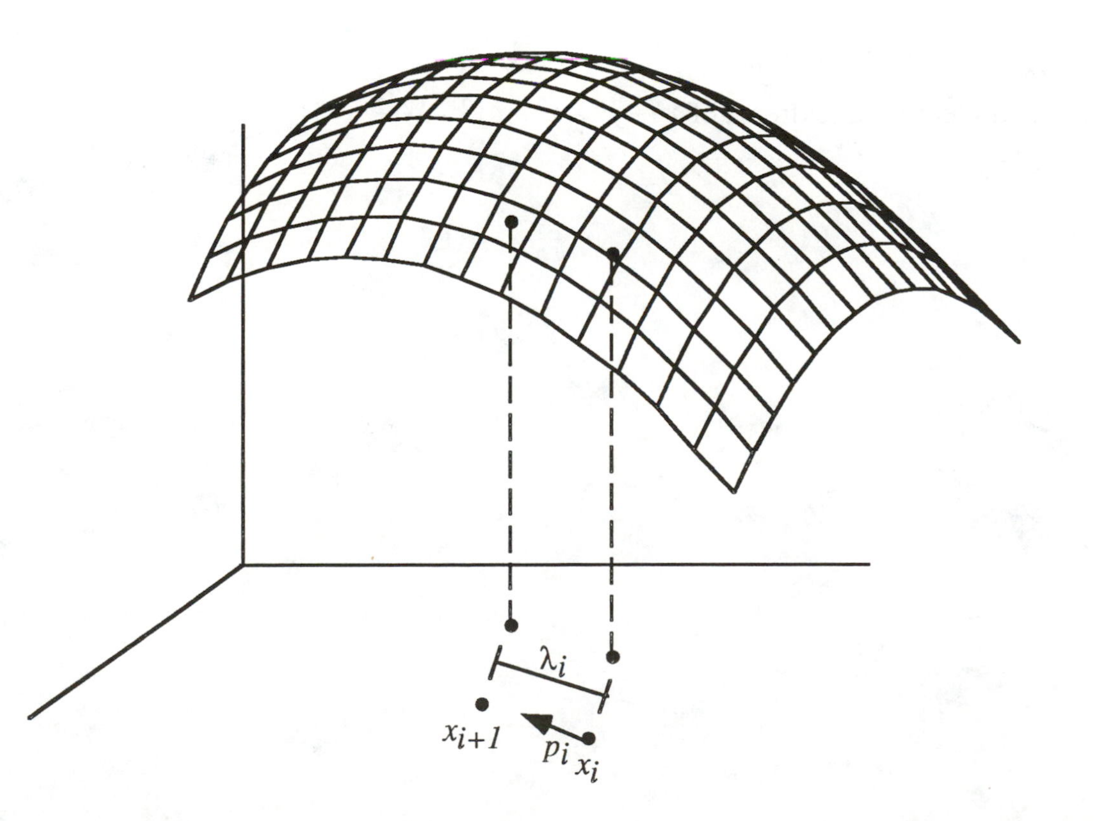

One of the oldest methods using this approach is Cauchy's method. [1847, *C.R. Academy of Science*, Paris, 25, pp 536-538] Based on the fact that the gradient points in the direction of the most rapid increase, Cauchy's method uses the gradient to determine the direction and then does a one dimensional optimization problem (a regular one variable max/min) in that direction to determine the step size. The one variable function can be obtained by parametrizing the curve along the surface in the direction of the gradient.

Part I: Find a book which discusses Cauchy's method and write down an algorithm which will perform the method easily. You will find discussions on Cauchy's method, also known as the "gradient method" or "steepest ascent method," in most introductory operations research or mathematical programming books.

Part II: Use your algorithm to do the following problems: (You can easily check your answers in the first two using ordinary multivariate max/min techniques in order to be certain you have created an algorithm that works.) In your report, include a discussion on what criterion you used for determining when to stop.

1. Find the minimum of $f(x,y) = 4x^2 + 4y^2 - 8x - 12y + 1$.

2. Find the maximum of $f(x,y) = 8x + 12y - 6x^2 - 8y^2 - 1$.

3. Find and identify all extrema of $x^4 + y^6 - x^2 - 4y^2 - x^2y^2$.

Information for the instructor only:

Problem abstract: Multivariate calculus students often fail to see the value of the gradient. Nonlinear optimization is an accessible, important application which can provide students a better understanding of the gradient as well as a deeper appreciation of its importance in other areas of mathematics and its applications. This project is intended to accomplish both of these goals. By discovering the algorithm themselves for specific examples (two of which can easily be visualized), students will need to see the gradient in the context of defining a curve along a given surface. This geometric construction should help them obtain ownership of the concept.

Prerequisite skills and knowledge: Students need to be familiar with multivariate calculus material, in particular, they will need to understand vectors and the gradient. Also, they will need to derive a parametrization of a curve along a surface in the direction of a given vector or do a Lagrange multiplier optimization.

Essential/useful library resources: Students will need access to introductory texts on mathematical programming or operations research in order to find a discussion of Cauchy's method.

Essential/useful computational resources: Programming in some form (Pascal, C, Maple, Mathematica, etc.) is necessary to avoid extensive computation.

Example of an acceptable approach: One possible algorithm follows:

Define f
Compute f_x and f_y
Set initial values x_0, y_0
** **unitgrad** $= \langle f_x(x_i,y_i), f_y(x_i,y_i)\rangle \dfrac{1}{\sqrt{f_x^2 + f_y^2}}$
$x(t) = x_i + t$ (**unitgrad**[1])
$y(t) = y_i + t$ (**unitgrad**[2])
λ = the value of t which minimizes/maximizes $f(x(t),y(t))$
$x_{i+1} = x_i + \lambda$ (**unitgrad**[1]), $y_{i+1} = y_i + \lambda$ (**unitgrad**[2])
print $x_{i+1}, y_{i+1}, f(x_{i+1}, y_{i+1})$
go to **.

Note: Rather than the one variable min/max described in this algorithm, one could do a Lagrange multiplier optimization on f with the gradient direction line as a constraint. This is not an easy problem and will require a very powerful tool to solve when f_x and f_y are complicated.

Solutions

(a) One iteration from any starting point (except the min., which has grad = 0) yields the minimum of this circular paraboloid: $x = 1, y = 1.5, f(1, 1.5) = -15$.

(b) Iterations from most initial points approach but never reach the maximum of this elliptical paraboloid. If one of the coordinates is correct in the initial point, the exact answer will be reached in one step:

$$x = \frac{2}{3}, y = \frac{3}{4}, f(\frac{2}{3}, \frac{3}{4}) = \frac{37}{6}.$$

(c) Not so easy to do by hand! (Do not expect students to find the exact values.)

The local max is at (0,0).

Four local minimums occur at $\left(\pm\sqrt{\frac{13+\sqrt{217}}{24}}, \pm\sqrt{\frac{1+\sqrt{217}}{12}}\right)$,

which are approximately $(\pm 1.07492, \pm 1.14495)$.

Four saddle points occur at $\left(0, \pm\sqrt[4]{\frac{4}{3}}\right)$ and $\left(\pm\sqrt{\frac{2}{2}}, 0\right)$,

which are approximately $(0, \pm 1.07457)$ and $(\pm 0.70711, 0)$.

Note: The first coordinate in the minimums and the nonzero coordinates in the saddle points are only so close by coincidence. They are really different.

References/bibliography/related topic: Here are a few of the many references for Cauchy's Method:

Winston, Wayne, *Operations Research, Applications and Algorithms*, Second Edition, PWS-Kent, 1991.

Taha, Hamdy, *Operations Research, An Introduction*, Fifth Edition, Macmillan, 1992.

Giordano and Weir, *A First Course in Mathematical Modeling*, Brooks/Cole, 1985.

Special implementation suggestions: It is important that students put together the Cauchy algorithm. A tendency is to have a computer do a monster Lagrange multiplier on the problem which is not the intent at all. In fact, it is hoped that 2(c) cannot be solved in this manner. An intermediate meeting with the instructor to verify that a reasonable algorithm has been obtained is strongly recommended. Also, students may be unable to determine how to parametrize a curve in the direction of the gradient unless they have seen similar computations in class. If this has not been done, a quick discussion with the group may be in order.

HISTORICAL PROJECTS

"The calculus ... is much more than a technical tool: it is a collection of abstract mathematical ideas which have accumulated over long periods of time. The foundations and central concepts are not today what they were in the seventeenth, eighteenth, or even in the nineteenth centuries and yet, in a sense, the unifying power and richness of its field of application depends not only on what the calculus is now but on all the concepts which have contributed in one way or another to its evolution ... Although there are many themes, some complementary and some contrasting, they all contain a common element of conflict, the conflict between the demands of mathematical rigor imposed by deductive logic and the essential nature of the infinitely great and the infinitely small perpetually leading to paradox and anomaly." (Baron, pp 1-3)

Following is a collection of projects designed to give students some exposure to how mathematics was done in the past: *Archimedes' Approximation of π, Archimedes' Determination of the Area of a Circle, Archimedes' Determination of the Surface Area of a Sphere, Cavalieri's Integration Method,* and *Zeno's Paradoxes* all foreshadowed the modern notions of limit, differentiation and/or integration. *Newton's Investigation of Cubic Curves* provides an interesting glimpse at some of the techniques used by Newton in the late seventeenth century. Archimedes wrote proofs which left out a number of details: like most mathematicians he did not record what he considered obvious. However, what was obvious to Archimedes is not necessarily obvious to a calculus student or even a mathematician today. Archimedes had a different and less sophisticated set of mathematical "tools" available to him, but he proved himself a master of the tools at his disposal. Key tools he did not possess were algebra and the modern notions of limit, differential or integral calculus, and the theory of infinite series. Today, we describe the numbers used in calculus as "real numbers," consisting of the union of two disjoint sets called the rational numbers (those that can be written as fractions) and irrational numbers (those that cannot be written as fractions). Archimedes did not have knowledge of set theory or the theory of real numbers, but he did make extensive use of Eudoxus' theory of proportions in much of his work. One example of his use of the theory of proportions is his contention that, given two circles, it must be true that the ratio of the circumference to the diameter of one is equal to the ratio of the circumference to the diameter of the other: $\frac{C_a}{d_a} = \frac{C_b}{d_b}$. We would make the equivalent statement that the circumference of any circle is equal to the diameter of the circle multiplied by the number π; $C = \pi d$, or $\frac{C}{d} = \pi$ for any circle.

These are some of the hardest projects in the volume: students are required to read and understand derivations and proofs which were built using mathematical tools that are either archaic or lacking in rigor by today's standards. Further, we are all challenged any time we attempt to follow the thoughts of great thinkers, and these projects give students the opportunity to carry on conversations with some of the great minds in the history of science and mathematics.

Zeno's paradoxes indicate that mathematicians of Archimedes' time had not resolved the question of whether they would consider time and magnitude to be infinitely divisible or made up of some smallest elements (infinitesimals). Archimedes, and many of his contemporaries, avoided the philosophical conflicts pointed out in Zeno's paradoxes by assuming that he could carry out certain geometrically-derived processes "as many times as necessary." Archimedes successfully applied that assumption in many of his impressive derivations and proofs. The guidelines for his approach are summarized below: (Baron, pages 34 & 35)

"1. Any finite quantity, however small, can be made as large as we please by multiplying it by a great enough number; or, given two unequal magnitudes a and b ($b < a$) there exists,

(i) a number n such that $nb > a$ (Euclid's *Elements*, Book V, definition 4),

(ii) a number n such that $n(a - b) > \gamma$, where γ is any magnitude whatsoever of the same kind (The Axiom of Archimedes, *On the Sphere and Cylinder*, Book I).

2. Any finite quantity can be made as small as we please by repeatedly subtracting from it a quantity greater than, or equal to, its half; or given two unequal quantities, a and b ($b < a$), there exists a number n, such that $a(1 - p)^n < b$, where $p > 1/2$ (Euclid's *Elements*, Book X, Proposition 1)."

Cavalieri, in the Middle Ages, tended to describe planar regions or solid objects as capable of being accurately described as a summation of a large number of very thin slices. This agrees with the intuitions of most beginning students, but does not address the philosophical conflicts pointed out in Zeno's paradoxes any better than Archimedes' approach. Cavalieri and even Newton and Liebniz approached their work with a tacit assumption of indivisibles rather than infinite divisibility. I find that for most students the limit notion (basically a nineteenth century refinement of the assumption of infinite divisibility) upon which calculus is based is less natural than either the assumption of the existence of indivisibles or the assumption that we can make quantities as small (or as large) as necessary. In this our students and our mathematical ancestors seem to agree.

Newton's Investigation of Cubic Curves is very different from the others in this section in that most of what is included in the primary project statement uses the same tools and ideas with which modern calculus students are comfortable. However, there are further explorations suggested that will lead students into the very unfamiliar but accessible area of affine changes in coordinates.

Special implementation suggestions:

You could assign these projects to a few small groups in a class. These topics could be the subject of a series of class discussions, in which each topic was briefly presented by the students who had done that particular project. In the end it would be useful for all students to seek to determine the common and unique insights of each of the different persons and/or methods and to pursue questions that were answered or raised by each of the different approaches. Hopefully they will gain a better understanding of the fundamental importance of the definitions of the derivative as the limit of a difference quotient and of the definite integral as the limit of a Riemann sum.

Consideration of our mathematical ancestors' efforts to deal with the very small and the very large might lead some students to be interested in studying non-standard analysis, as developed by the American mathematician Abraham Robinson and others, but that's another project, probably as part of a later course.

Historical Projects Bibliography:

Aaboe, Asger: *Episodes from the Early History of Mathematics*, Random House, 1964.
Ball, W. W. R.: *Newton's classification of cubic curves, Proceedings of the London Math. Society* 22, 104-143, 1890.
Baron, Margaret E.: *The Origins of the Infinitesimal Calculus*, Pergamon Press, 1969.
Boyer, C.B.: *The Concepts of Calculus*, New York, 1949.
Boyer, C.B. and Merzbach, U.C.: *A History of Mathematics*, Second Edition, Wiley, 1989.
Brieskorn, Edgert and Knörrer, Horst: *Plane Algebraic Curves*, Birkhäuser, Boston, 1986.
Cajori, F.P.: *The History of Zeno's Arguments on Motion*, published as a series of articles in *The American Mathematical Monthly*, Volume XXII, Numbers 1-7, 1915.
Courant, R.: *Differential and Integral Calculus*, Volume I, Second Edition.
Dickson, L. E.: *New First Course in the Theory of Equations*, Wiley, New York, 1939.
Dijksterhuis, E.J.: *Archimedes*, The Humanities Press, 1957.
Edwards, C.H., Jr.: *The Historical Development of the Calculus,* Springer-Verlag, 1979.
Euclid: *Elements.*
Evans, G.W.: Cavalieri's theorem in his own words, *American Mathematical Monthly* 24, pp 447-451, 1917.
Eves, H.: Two surprising theorems on Cavalieri congruence, *The College Mathematics Journal*, 22:2, pp 118-124, March 1991.
Fauvel, J. and J. Gray, eds.: *The History of Mathematics, A Reader*, Macmillan, 1987.
Heath, T.L.: *A Manual of Greek Mathematics*, Dover, 1963.
Heath, T.L.: *The Works of Archimedes*, Cambridge University Press, 1897.
Henle, J.M. and Kleinberg, E.M.: *Infinitesimal Calculus*, The MIT Press, 1979.
Hofstadter, D.: *Godel, Escher, Bach: An Eternal Golden Braid*, Basic Books.
Hooper, A.: *Makers of Mathematics*, Chapter VI, Random House, 1948.
Hurd and Loeb: *Nonstandard Analysis: An Introduction to Nonstandard Real Analysis*, Academic Press, 1985.
Kline, M.: *Mathematical Thought from Ancient to Modern Times*, Oxford University Press, 1972.
Newton, Isaac: *The Mathematical Works of Isaac Newton*, edited by Derek Whiteside, vol. 2 and 7, Cambridge University Press, 1967-1981.
Primrose, E. J. F.: *Algebraic Plane Curves*, Macmillan, London, 1955.
Russell, B.: "The Problem of Infinity Considered Historically," in *Our Knowledge of the External World.*
Salmon, George: *A Treatise on the Higher Plane Curves*, Hodges, Forster, and Co., Dublin, 1873.
Salmon, W.C., ed.: *Zeno's Paradoxes*, Bobbs-Merrill, 1970.
Simmons, George: *Calculus with Analytic Geometry*, McGraw Hill, New York, 1985.
Stillwell, J.: *Mathematics and its History*, Springer-Verlag, 1989.
Stillwell, John: *Mathematics and Its History*, Springer-Verlag, New York, 1989.
Struik, D.J., editor: *A Source Book in Mathematics, 1200-1800*, Princeton University Press, 1986.
Toeplitz, O.: *The Calculus: A Genetic Approach*, Berlin, 1949 (translated by L. Lange, Chicago, 1963).

Title: Archimedes' Determination of the Area of a Circle

Authors: Mic Jackson and Sarah Angley, Earlham College

Problem Statement: Archimedes wrote proofs which left out a number of details: like most mathematicians he did not record what he considered obvious. However, what was obvious to Archimedes is not necessarily obvious to a mathematician today. Inclusion of some additional details would have made his works much easier for us to follow. Archimedes had a different and less sophisticated set of mathematical "tools" available to him, but he proved himself a master of the tools at his disposal. In this particular proof, he used the assumption that processes could be carried on indefinitely (his way of avoiding the notion of infinity), theorems from Euclidean geometry, and a form of proof known as *reductio ad absurdum.* Key tools he did not possess were algebra, modern notions of limit, differential or integral calculus, and the theory of infinite series.

Your task is to justify the 16 statements highlighted in **bold** face in Archimedes' proof that the area of a circle is equal to πr^2. To justify in this case means to provide the details that Archimedes left out so that the proof is more easily readable. Show how each statement follows from known definitions, axioms and/or theorems and from previous statements in this proof. You will need to make geometrical arguments and use some of your knowledge of infinite series to understand Archimedes. In each case, be sure you explain what Archimedes meant and why he felt the phrase was necessary .

Replication of Archimedes' determination of the area of a circle.
(Adapted from "Measurement of a Circle," *The Works of Archimedes* , by T.L. Heath)

Proposition 1: The area of a circle is equal to the area of a right triangle in which one of the legs has the same length as the radius of the circle and the other leg has the same length as the circumference.

Let circle O be the given circle and K the area of the triangle described. (Note that the triangle with area K shown in Figure 1 is not to scale.)

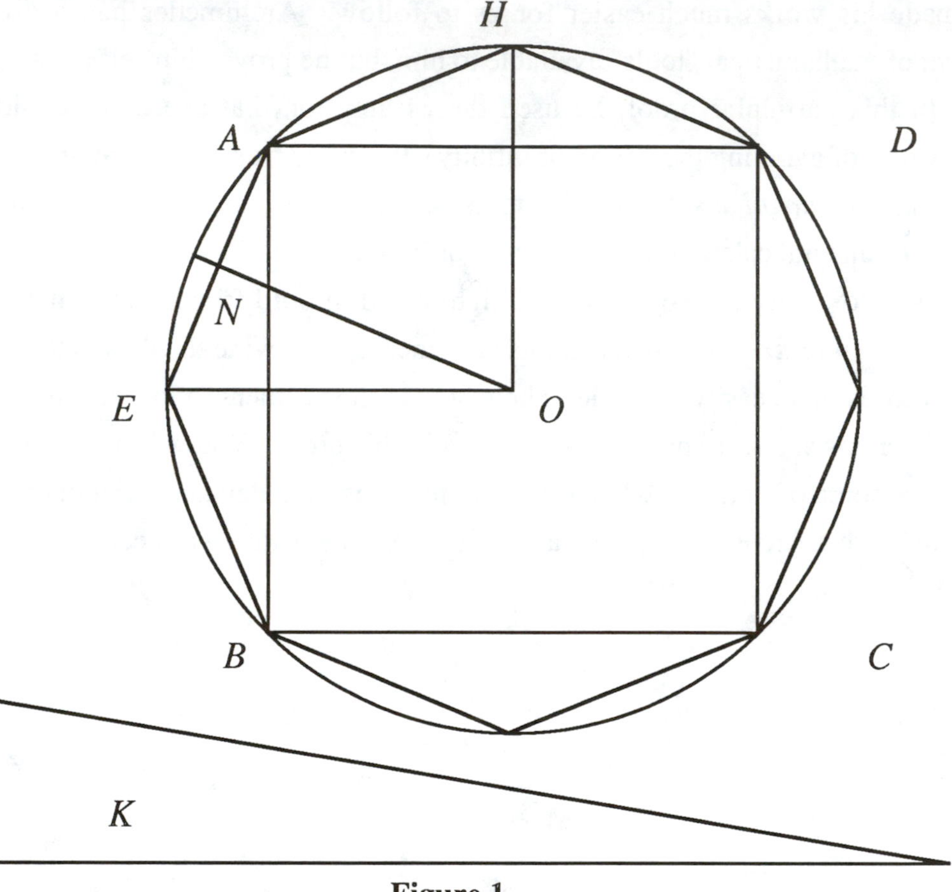

Figure 1

Then, **(*) if the area of circle *O* is not equal to *K*, it must be either greater than *K* or less than *K*.** (*Example of an acceptable justification: The Trichotomy Property states that if two quantities are compared, then either they are equal or one must be greater than the other.)

Case I Assume that the area of circle O is greater than K.

Inscribe a square $ABCD$ inside circle O (see Figure 1). Bisect the arcs AB, BC, CD and DA, and connect the points on circle O with straight line segments to form a regular octagon. Then bisect (if necessary) these eight arcs,

(1) and continue this until the sides of the inscribed regular polygon subtend segments of the circle, the sum of whose areas is less than the difference of the area of circle

O and the area K.

(2) We can conclude that the area of the inscribed regular polygon is greater than K.

Now, let AE be any side of the inscribed regular polygon, and ON the perpendicular drawn to AE from the center O. Then

(3) ON is less than the radius of circle O and therefore ON is less than one of the legs of the given right triangle with area K.

(4) Also the perimeter of the inscribed regular polygon is less than the circumference of the circle. Therefore, the perimeter of the inscribed regular polygon is less than the other leg of the triangle with area K.

(5) Therefore the area of the inscribed polygon is less than K.

(6) Our hypothesis has led to a contradiction.

(7) Thus the area of circle O is not greater than K.

Case II Assume that the area of circle O is less than K.

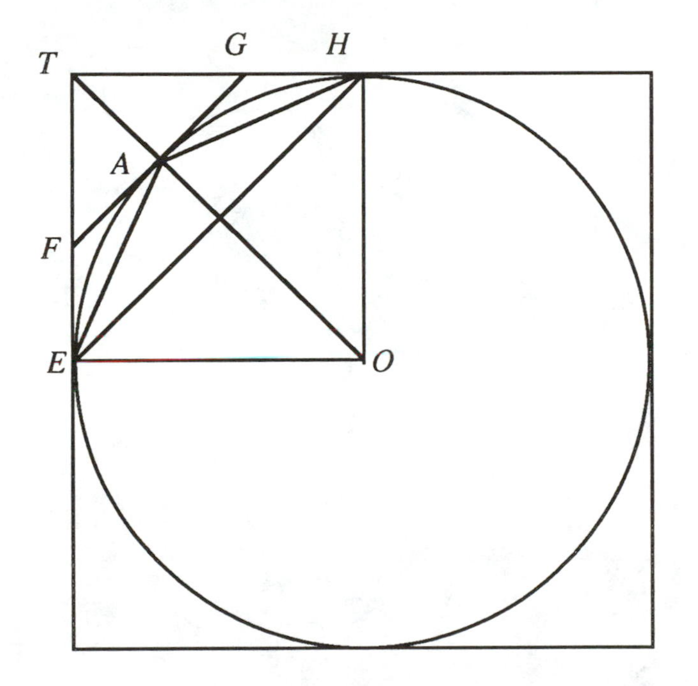

Circumscribe a square about circle O, and let T be the vertex of the square where the two adjacent sides which touch circle O at points E and H intersect. Let A be the point where the bisector of arc EH intersects arc EH and draw the tangent line to circle O at point A. Notice that line OA passes through point T. Let F and G be the points where the tangent line through A intersects the square.

Then

(8) the angle TAG is a right angle.

(9) Therefore, $TG > GA$ and $TG > GH$.

(10) It follows that the area of triangle FTG is greater than half the area of region $TEAH$ (where EAH is the path along the arc of circle O).

(11) Similarly, if we bisect the arc AH and draw the tangent at the point of bisection,

that tangent will cut off more than one-half of the area of region *GAH*.

(12) If we continue this process, we will eventually arrive at a circumscribed regular polygon for which the spaces intercepted between it and the circle are together less than the difference of area K and the area of circle *O*.

(13) Thus the area of that circumscribed regular polygon will be less than *K*. Now, since the perpendicular from *O* on any side of the circumscribed regular polygon is equal to the radius of the circle, while the perimeter of that polygon is greater than the circumference of the circle,

(14) it follows that the area of the circumscribed regular polygon is greater than *K*.

(15) Therefore the area of the circle is not less than *K*.

(16) Since then the area of the circle is neither greater nor less than *K*, it is equal to it.

Information for the instructor only:

Problem abstract: Archimedes' determination that the area of a circle was equal to that of a right triangle with height the radius of the circle and base the circumference of the circle is a classic mathematical work. His work here provides an excellent example of a *reductio ad absurdum* proof, the standard form of rigorous proof used by mathematicians from the time of the Greeks to the 17th Century. His work also involves carrying out a process "as many times as necessary," a sort of informal passing to the limit. Although the result is well known, attempting to replicate Archimedes' thoughts will prove a challenge to the reading and mathematical skills of all students.

Prerequisite skills and knowledge: high school geometry, form of *reductio ad absurdum* proofs.

Essential/useful library resources: See the bibliography in the introduction to Historical Projects.

Essential/useful computational resources: none

Example of an acceptable approach:

(1) and continue this until the sides of the inscribed regular polygon subtend segments of the circle, the sum of whose areas is less than the difference of the area of circle O and the area K.

(Keep up the process of bisecting arcs, yielding an 8-gon, a 16-gon, a 32-gon,... The area of these regular n-gons keeps increasing, but the incremental increase is getting smaller. Although Archimedes did not have our understanding of convergent infinite series, he did postulate that the area of any inscribed polygon must be less than the area of the circle; and since the polygons were increasing in area, he reasoned that the increase must be getting arbitrarily small as n increased. In effect the n-gon "fills" more of the circle as n increases.)

(2) We can conclude that the area of the inscribed regular polygon is greater than K.

(The difference in the areas of the circle and polygon is less than the difference in the areas of the circle and the triangle. $[(c-p) < (c-t)] \rightarrow p > t$)

(3) ON is less than the radius of circle O

(it doesn't reach the circle).

(4) perimeter of the inscribed regular polygon is less than the circumference of the circle

(the straight line is the shortest distance between two points, so each edge of the polygon has length less than the corresponding arc of the circle).

(5) Therefore the area of the inscribed polygon is less than K.

(Each side of the polygon is the base of an isosceles triangle with height ON and the polygon is made up of "$2n$" of these triangles. The area of the polygon is the sum of the areas of these $2n$ triangles and is equal to the area of a triangle with height ON and base the perimeter of the polygon. The area of this triangle is clearly less than the area, K, of a triangle with base the perimeter of the circle and height the radius of the circle.)

(6) Our hypothesis has led to a contradiction.

(The hypothesis resulted in two incompatible conclusions, that the area of the polygon is both less than and greater than K.)

(7) Thus the area of circle O is not greater than K.

(Making the assumption that the area of the circle is greater than K led to a contradiction.)

(8) the angle TAG is a right angle.

(Tangent lines are perpendicular to the radius at the point of tangency, and line segment OA is the radius to the circle through point A.)

(9) Therefore, $TG > GA$ and $TG > GH$.

($TG > GA$ because TG is the hypotenuse of the right triangle TAG, while GA is a leg of that same triangle. Now, construct segment GO, note that it equals itself. Note also that segment GH is tangent to the circle at point H, so angle GHO is a right angle. Segments AO and HO are equal, being radii of the circle. Hence, triangle GOA is congruent to triangle GOH. So $GA = GH$ because they are corresponding parts of congruent triangles. Finally, $TG > GH$.)

(10) It follows that the area of triangle FTG is greater than half the area of region $TEAH$.

(Consider triangles TAG and GAH. Both have the same height, the perpendicular distance from point A to line TH; but TAG has a larger base than GAH, so the area of triangle TAG is greater than the area of triangle GAH. If we double the area of TAG, we get exactly the area of triangle FTG. And if we add the area of TAG to the area of GAH, we get one-half the area of $TEAH$. Having added equal quantities to both sides of an inequality, the inequality remains: The area of $FTG > 1/2$ the area of polygon $TEAH$ and hence $1/2$ of the area of the desired region $TEAH$.)

(11) Similarly, if we bisect the arc AH and draw the tangent at the point of bisection, that tangent will cut off more than one-half of the area of region GAH.

(We can go through an argument identical to that made for items 8, 9 and 10.)

(12) If we continue this process, we will eventually arrive at a circumscribed regular polygon for which the spaces intercepted between it and the circle are together less than the difference of area K and the area of circle O.

(In Figure 2 below, notice that the initial error consists of 4 regions identical to the region $TEAH$, which resembles a triangle with a curved base. After completing the construction described above, the total error in approximation consists of 8 regions identical to region GAH, another "triangle" with a curved base.

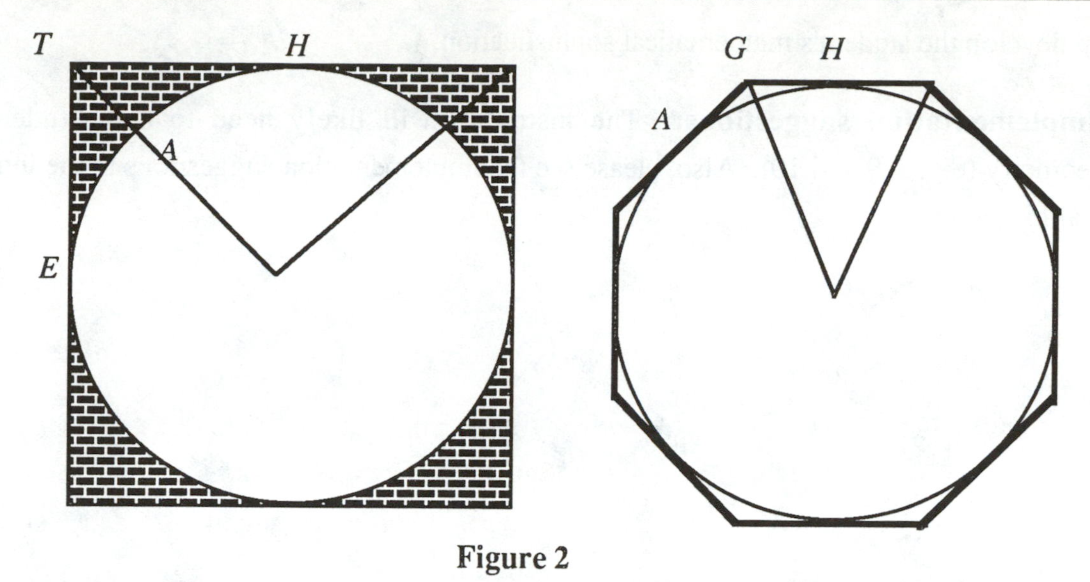

Figure 2

Since the error in the new approximation is less than half the previous error, Archimedes knew by axiom that he could repeat the construction as described until the error of approximation was as small as he wanted/needed it to be. (Today we might say something like "If $F(x) < \frac{x}{2}$, then for any $\varepsilon > 0$, then there exists a counting number N such that if $n > N$ then $F^{(n)}(x) < \varepsilon$.") Notice that our first polygonal approximation of a circle was a square (4 sides), the second a regular octagon because we bisected each of four equal arcs. The next polygonal approximation will be formed by bisecting each of eight equal arcs, and constructing tangents at each point where the bisector meets the circle. The vertices of the resulting regular 16-gon will be the points where the tangent lines intersect. And so on...)

(13) Thus the area of that circumscribed regular polygon will be less than K.
(It is closer in area to the circle than K.)

(14) It follows that the area of the circumscribed regular polygon is greater than K.
(As before, K is the area of the triangle with base the perimeter of the circle and height the radius of the circle. The area of the polygon is equal to the area of a triangle with the same height and a larger base. So the area of the polygon is larger than K.)

(15) Therefore the area of the circle is not less than K.
(The assumption that the area of the circle is greater than K lead to two contradictory conclusions.)

(16) Since then the area of the circle is neither greater nor less than K, it is equal to it. (Application of the Trichotomy Property.)

References/bibliography/related topics:

(1) "On the Sphere and the Cylinder," "The Sand Reckoner," "Quadrature of the Parabola," all by Archimedes, and "Anticipations by Archimedes of the Integral Calculus;" in Heath's, *The Works of Archimedes*.

(2) A number of the sections in "Greek Mathematics," Chapter 1, in Baron's book are interesting and

would help develop the student's mathematical sophistication.

Special implementation suggestions: The instructor will likely need to help students recall relevant geometry (e.g. at 9 and 10). Also, please see the implementation suggestions in the introduction to Historical Projects.

Title: Archimedes' Approximation of π

Authors: Mic Jackson and David May, Earlham College

Problem Statement: In this classic work, Archimedes used the "compression method," where he showed that the actual value of the ratio of the circumference of a circle to its diameter (π) must be less than a certain geometrically derived sequence of values and greater than another geometrically derived sequence of values. By doing so, he was able to come up with an approximation of π that was the best until relatively recent times. Archimedes also used the assumption that these geometrical processes could be carried on indefinitely (his way of dealing with the notion of infinity), Eudoxes' theory of proportions (his way of dealing with the problem of the existence of irrational numbers), and a form of proof known as *reductio ad absurdum*. Anyone attempting to follow Archimedes' work needs to be competent with the rules and concepts of proportion which were characteristic of ancient Greek mathematics. Some are listed below.

(1) If $\frac{a}{b}=\frac{b}{c}$, then $\frac{a}{c}=\frac{a^2}{b^2}$. **(2)** If $\frac{a}{b}=\frac{b}{c}=\frac{c}{d}$, then $\frac{a}{d}=\frac{a^3}{b^3}$.

(3) If $\frac{a}{b}=\frac{m}{n}$ and $\frac{b}{c}=\frac{p}{q}$, then $\frac{a}{c}=\frac{mp}{nq}$.

(4) If $\frac{a}{b}=\frac{c}{d}$, then (a) $\frac{a}{c}=\frac{b}{d}$, (b) $\frac{b}{a}=\frac{d}{c}$, (c) $\frac{a+b}{b}=\frac{c+d}{d}$,

(d) $\frac{a-b}{b}=\frac{c-d}{d}$ if $a>b$, (e) $\frac{a}{a-b}=\frac{c}{c-d}$ if $a>b$,

(f) $\frac{a+b}{a}=\frac{c+d}{c}$, (g) $\frac{b}{a-b}=\frac{d}{c-d}$, (h) $\frac{a-b}{a}=\frac{c-d}{c}$.

(5) If $\frac{a}{b}>\frac{c}{d}$, then (a) $\frac{a}{c}>\frac{b}{d}$, (b) $\frac{b}{a}<\frac{d}{c}$, (c) $\frac{a+b}{b}>\frac{c+d}{d}$,

(d) $\frac{a-b}{b}>\frac{c-d}{d}$, (e) $\frac{a}{a-b}<\frac{c}{c-d}$, (f) $\frac{a+b}{a}<\frac{c+d}{c}$,

(g) $\frac{b}{a-b}<\frac{d}{c-d}$, (h) $\frac{a-b}{a}>\frac{c-d}{c}$; (i) if also $c>d$, then $a>b$.

(6) If $a>b$, and c is "any magnitude homogeneous with a and b," then $\frac{a}{b}>\frac{a+c}{b+c}$.

Your task is three-fold:

(1) Justify the numbered statements in the following work. Show how each statement follows from known definitions, axioms and/or theorems and from previous statements. You will need to make use of your knowledge of geometry, proportions and infinite sequences.

(2) Extend Archimedes' approximation one step further by finding the upper and lower bounds for π using regular inscribed and circumscribed 192-gons. You may use decimal approximations in your calculations if you are uncomfortable working with fractions (Archimedes had no choice).

(3) You should also consider the question of how his work foreshadows our current notion of limit.

Proposition: The ratio of the circumference of any circle to its diameter is less than $3\frac{1}{7}$ but greater than $3\frac{10}{71}$.

(Adapted from *The Works of Archimedes*, by T.L. Heath, pages 93 to 98)

Part I

Let segment AB be the diameter of circle O, let segment AC be tangent to circle O at point A, and let $\angle AOC$ be $\frac{1}{3}^{\text{rd}}$ of a right angle. Then it follows that:

(1) $\dfrac{OA}{AC} = \dfrac{\sqrt{3}}{1} > \dfrac{265}{153}$ and (2) $\dfrac{OC}{CA} = \dfrac{2}{1} = \dfrac{306}{153}$.

[Hint: Archimedes used a rather complicated method based on continued fractions to compute his approximations of $\sqrt{3}$. You should show the validity of item (1) using trigonometry, but you are welcome to consult with your instructor if you want to see how Archimedes did it.]

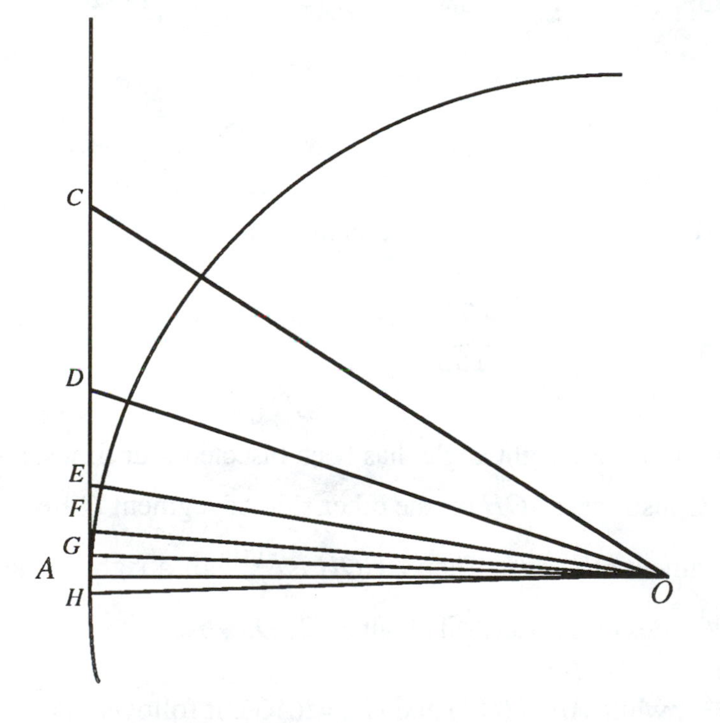

First, draw segment OD bisecting $\angle AOC$ and meeting segment AC in point D. Now,

(3) $\dfrac{CO}{OA} = \dfrac{CD}{DA}$ [Hint: This result is Proposition 3 in the 6th book of Euclid's *Elements*. You should look it up and be sure you understand his proof.]

(4) $\dfrac{CO+OA}{OA} = \dfrac{CA}{DA}$ (5) $\dfrac{CO+OA}{CA} = \dfrac{OA}{AD}$ (6) $\dfrac{OA}{AD} > \dfrac{571}{153}$

(7) $\dfrac{OA^2+AD^2}{AD^2} > \dfrac{571^2+153^2}{153^2}$ (8) $\dfrac{OD^2}{AD^2} > \dfrac{349450}{23409}$ (9) $\dfrac{OD}{DA} > \dfrac{591\frac{1}{8}}{153}$

Second, let segment OE bisect $\angle AOD$, meeting segment AD in point E.

(10) $\dfrac{DO}{OA} = \dfrac{DE}{EA}$ (11) $\dfrac{DO+OA}{DA} = \dfrac{OA}{AE}$

(12) $\dfrac{OA}{AE} > \dfrac{1162\frac{1}{8}}{153}$ (13) $\dfrac{OE^2}{EA^2} > \dfrac{\left(1162\frac{1}{8}\right)^2 + 153^2}{153^2} = \dfrac{1373943\frac{33}{64}}{23409}$

(14) $\dfrac{OE}{EA} > \dfrac{1172\frac{1}{8}}{153}$

Third, let segment OF bisect $\angle AOE$ and meet segment AE in point F.

(15) $\dfrac{OA}{AF} > \dfrac{1162\frac{1}{8} + 1172\frac{1}{8}}{153} = \dfrac{2334\frac{1}{4}}{153}$ (16) $\dfrac{OF^2}{FA^2} > \dfrac{\left(2334\frac{1}{4}\right)^2 + 153^2}{153^2} = \dfrac{5472132\frac{1}{16}}{23409}$

(17) $\dfrac{OF}{FA} > \dfrac{2339\frac{1}{4}}{153}$

Fourth, let segment OG bisect $\angle AOF$, meeting segment AF in point G.

(18) $\dfrac{OA}{AG} > \dfrac{2334\frac{1}{4} + 2339\frac{1}{4}}{153} = \dfrac{4673\frac{1}{2}}{153}$

Now $\angle AOC$, which is one-third of a right angle, has been bisected four times, and it follows that $\angle AOG$ is $\frac{1}{48}$th of a right angle. Construct $\angle AOH$ on the other side of segment OA equal to $\angle AOG$, and let the ray GA produced meet segment OH in H. Then $\angle GOH$ is $\frac{1}{24}$th of a right angle. Thus GH is one side of a regular polygon of 96 sides circumscribed about circle O.

And, since $\dfrac{OA}{AG} > \dfrac{4673\frac{1}{2}}{153}$, while $AB=2(OA)$ and $GH=2(AG)$, it follows that

(19) $\dfrac{AB}{\text{perimeter of 96-gon}} > \dfrac{4673\frac{1}{2}}{(96)(153)} = \dfrac{4673\frac{1}{2}}{14688}$.

(20) But $\dfrac{14688}{4673\frac{1}{2}} = 3 + \dfrac{667\frac{1}{2}}{4673\frac{1}{2}} < 3 + \dfrac{667\frac{1}{2}}{4672\frac{1}{2}} = 3\frac{1}{7}$.

Therefore the circumference of the circle (being less than the perimeter of the polygon) is less than $3\frac{1}{7}$ times the diameter AB.

Part II

Let segment AB be the diameter of circle O, and let segment AC which meets circle O in point C make $\angle CAB$ equal to one-third of a right angle. Construct segment BC. Then

(1) $\dfrac{AC}{CB} = \dfrac{\sqrt{3}}{1} < \dfrac{1351}{780}$ and (2) $\dfrac{BA}{BC} = \dfrac{2}{1} = \dfrac{1560}{780}$.

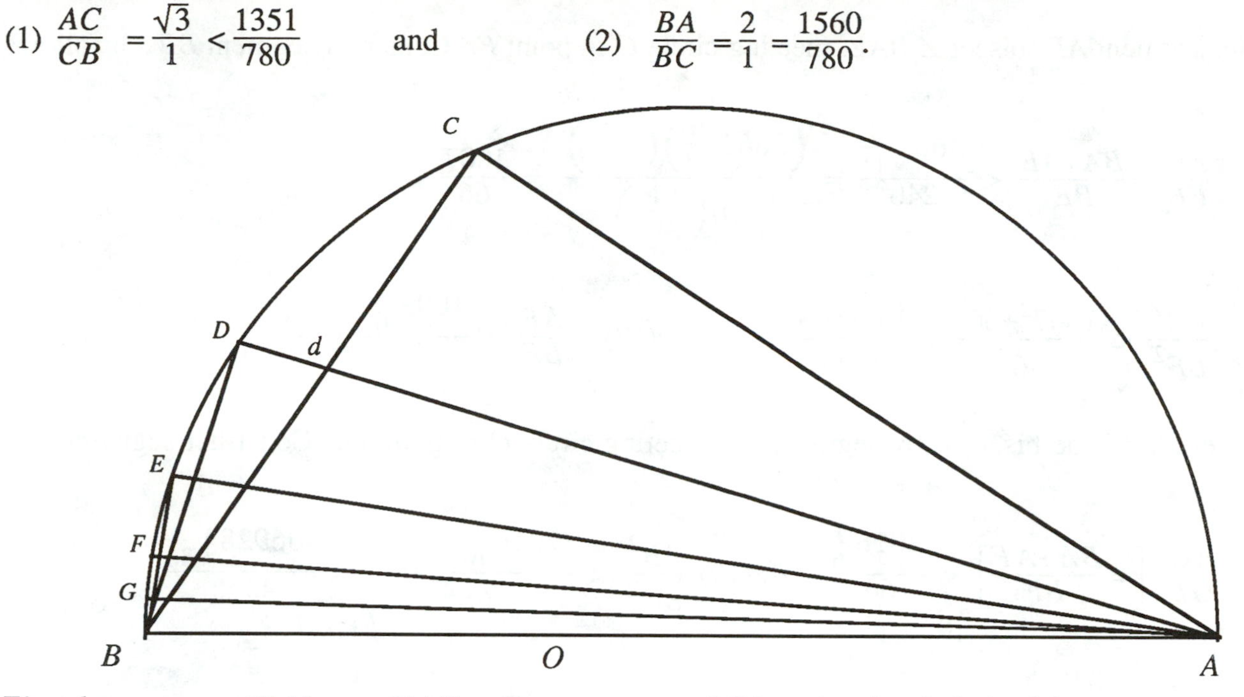

First, let segment AD bisect $\angle BAC$ and meet segment BC in point d and circle O in point D. Construct segment BD. Then

(3) $\angle BAD = \angle dAC = \angle dBD$

(4) $\angle BCA$ and $\angle BDA$ are both right angles.

(5) $\Delta ADB \sim \Delta ACd \sim \Delta BDd$

(6) $\dfrac{AD}{DB} = \dfrac{BD}{Dd} = \dfrac{AC}{Cd}$

(7) It is also true that $\dfrac{AC}{Cd} = \dfrac{AB}{Bd}$. [This has the same justification as item (3) in Part I.]

(8) $\dfrac{AB+AC}{Bd+Cd} = \dfrac{BA+AC}{BC} = \dfrac{AD}{DB}$

(9) $\dfrac{AD}{DB} < \dfrac{1560+1351}{780} = \dfrac{2911}{780}$

(10) $\dfrac{AB^2}{BD^2} < \dfrac{2911^2+780^2}{780^2} = \dfrac{9082321}{608400}$

(11) and $\dfrac{AB}{BD} < \dfrac{3013\frac{3}{4}}{780}$

Second, let segment AE bisect $\angle BAD$, meeting circle O in point E; and construct segment BE. Then we prove, in the same way as before, that

(12) $\dfrac{AE}{EB} = \dfrac{BA+AD}{BD} < \dfrac{3013\frac{3}{4} + 2911}{780} = \dfrac{5924\frac{3}{4}}{780} = \dfrac{\left(5924\frac{3}{4}\right)\left(\frac{4}{13}\right)}{(780)\left(\frac{4}{13}\right)} = \dfrac{1823}{240}$.

(13) $\dfrac{AB^2}{BE^2} < \dfrac{1823^2+240^2}{240^2} = \dfrac{3380929}{57600}$ (14) $\dfrac{AB}{BE} < \dfrac{1838\frac{9}{11}}{240}$.

Third, let segment AF bisect $\angle BAE$, meeting circle O in point F. Construct segment BF. It follows that

(15) $\dfrac{AF}{FB} = \dfrac{BA+AE}{BE} < \dfrac{3661\frac{9}{11}}{240} = \dfrac{\left(3661\frac{9}{11}\right)\left(\frac{11}{40}\right)}{(240)\left(\frac{11}{40}\right)} = \dfrac{1007}{66}$

(16) $\dfrac{AB^2}{BF^2} < \dfrac{1007^2+66^2}{66^2} = \dfrac{1018405}{4356}$ (17) $\dfrac{AB}{BF} < \dfrac{1009\frac{1}{6}}{66}$.

Fourth, let $\angle BAF$ be bisected by segment AG meeting circle O in point G. Construct segment GB. Then

(18) $\dfrac{AG}{GB} \left[= \dfrac{BA+AF}{BF}\right] < \dfrac{2016\frac{1}{6}}{66}$ (19) $\dfrac{AB^2}{BG^2} < \dfrac{(2016\frac{1}{6})^2+66^2}{66^2} = \dfrac{4069284\frac{1}{36}}{4356}$

(20) $\dfrac{AB}{BG} < \dfrac{2017\frac{1}{4}}{66}$ (21) $\dfrac{BG}{AB} > \dfrac{66}{2017\frac{1}{4}}$

Note that because $\angle BAG$ is the result of the fourth bisection of $\angle BAC$ it is equal to $\frac{1}{48}$th of a right angle. Thus the angle subtended by BG at the center of circle O is $\frac{1}{24}$th of a right angle. Therefore BG is a side of a regular inscribed polygon of 96 sides. It follows from (21) that

(22) $\dfrac{\text{perimeter of 96-gon}}{AB} > \dfrac{6336}{2017\frac{1}{4}}$ and (23) $\dfrac{6336}{2017\frac{1}{4}} > 3\frac{10}{71}$.

Therefore the circumference of circle O must be greater than $3\frac{10}{71}$ times its diameter.

(24) Thus $3\frac{10}{71} < \dfrac{\text{circum}}{\text{diam}} < 3\frac{1}{7}$.

Information for the instructor only:

Problem abstract: Archimedes' approximation of π using inscribed and circumscribed regular polygons is a classic mathematical work accessible to typical calculus students. His work here provides an excellent example of the way that Greek mathematicians dealt with what we might call the limit notion. His approximation involves carrying out a process "as many times as necessary," providing a sort of informal passing to the limit. Although we have much better approximations of π, his was one of the best until relatively recent times. In addition, attempting to replicate Archimedes' thoughts and calculations will provide a challenge to the reading and mathematical skills of all students.

Prerequisite skills and knowledge: high school geometry and algebra, including the theory of proportions.

Essential/useful library resources: See the bibliography in the introduction to Historical Projects.

Essential/useful computational resources: All of the calculations can be done by hand or with a typical scientific calculator.

Example of an acceptable approach: I have not included an example of the mathematical justifications students would need to make because most of the details Archimedes left out have been supplied; the students will need to see and describe the connections between steps. However, I have included some important notions in the **References and related resources** section. Below is an acceptable example of their extension of Archimedes' approximation, using decimal approximations.

I. Circumscribed

Fifth, let OP bisect angle AOG. Then $\frac{OA}{AP} = \frac{GO+AO}{AG} > \frac{9349.503769}{153}$.

Note that $\frac{OG^2}{GA^2} = \frac{OA^2+AG^2}{GA^2} > \frac{\left(4673\,\frac{1}{2}\right)^2+153^2}{153^2} = \frac{21865011\,\frac{1}{4}}{23409}$, so $\frac{OG}{GA} > \frac{4676.003769}{153}$

and $\frac{GO+AO}{GA} > \frac{4676.003769 + 4673.5}{153}$.

So, angle AOP is the result of the fifth bisection of the angle AOC and is equal to $\frac{1}{96}^{\text{th}}$ of a right angle. Now make angle AOQ = angle AOP on the other side of OA. Then angle $POQ = \frac{1}{48}^{\text{th}}$ of a right angle and PQ is one side of a regular 192-gon circumscribed about circle O.

Since $\frac{OA}{AP} > \frac{9349.503769}{153}$ and $\frac{AB}{PQ} = \frac{OA}{AP}$, we have

$$\frac{AB}{\text{perimeter of the 192-gon}} > \frac{9349.503769}{(153)(192)} \quad \text{or} \quad \frac{\text{perimeter of the 192-gon}}{AB} < 3.141984936.$$

II. Inscribed

Fifth, let AH bisect angle BAG, and construct BH.

$$\text{Then } \frac{AH}{HB} = \frac{BA+AG}{BG} < \frac{2017\frac{1}{4} + 2016\frac{1}{6}}{66} = \frac{4033.416667}{66} \approx 61.11237374.$$

$$\text{Hence } \frac{AB^2}{BH^2} = \frac{AH^2+BH^2}{BH^2} < \frac{\left(2017\frac{1}{4} + 2016\frac{1}{6}\right)^2 + 66^2}{66^2} \approx 3735.722224.$$

So $\frac{AB}{BH} < 61.12055484$, and $\frac{BH}{AB} > 0.01636110802$. Since angle BAH is the result of the fifth bisection of angle BAC it is $\frac{1}{96}$th of a right angle; and angle BOH is $\frac{1}{48}$th of a right angle. So BH is a side of a regular 192-gon.

$$\text{Hence } \frac{\text{perimeter of the 192-gon}}{AB} > (192)(0.01636110802) \approx 3.14133274.$$

$$\text{So } 3.14133274 < \frac{C}{d} < 3.141984936.$$

Note that this is a slight (one decimal place) improvement on Archimedes' result:

$$3.1408450704 < \frac{C}{d} < 3.1428571429.$$

Conjectures we expect that some students will make: Perhaps magnitudes *are* made up of infinitesimal elements.

Questions for further exploration:

(1) Archimedes' made a number of other significant determinations, most of which can be found in the library resources listed. A bright student would gain from tackling one on her own. Most students could gain from reading about Archimedes' and his accomplishments. Archimedes did not talk about limits in the modern sense, but his notion of repeating a process an indefinite number of times foreshadows modern limit theory. An eager student would profit from reading more about this topic.

(2) A calculus founded on the notion of infinitesimals, rather than accepted limit theory, known as "non-standard analysis," (see Henle and Kleinberg) could provide the unusually bright and motivated student with some food for thought.

(3) The history of the calculation of π is filled with new, strange and/or interesting mathematics; a useful study for a good student. She could start with *An Introduction to the History of Mathematics*,

fifth edition, by Howard Eves, pages 85-89, 96.

References/bibliography/related topics:

(1) "On the Sphere and the Cylinder," "The Sand Reckoner," "Quadrature of the Parabola," all by Archimedes, and "Anticipations by Archimedes of the Integral Calculus;" in Heath's, *The Works of Archimedes*.

(2) A number of the sections in "Greek Mathematics," Chapter 1, in Baron's book are interesting and would help develop the student's mathematical sophistication.

(3) A lengthy series of articles entitled "The History of Zeno's Arguments on Motion," by Florian Cajori, the American Mathematical Monthly published during the year of 1915 gives thorough coverage of the philosophical problems related to the very small and the very large.

(4) There is an interesting discussion of the number π as a limit in *Differential and Integral Calculus*, Volume I, Second Edition, by R. Courant, pages 44-46. This brief article begins: "A limiting process which in essence goes back to classical antiquity is that by which the number π is defined. Geometrically π means the area of the circle of radius 1. We therefore accept the existence of this number π as intuitive, regarding it as obvious that this area can be expressed by a (rational or irrational) number, which we then simply denote by π. However, this definition is not of much help to us if we wish to calculate the number with any accuracy. We have then no choice but to represent the number by means of a limiting process, namely, as the limit of a sequence of known and easily calculated numbers. Archimedes himself used this process in his method of exhaustions, where he steadily approximated to the circle by means of regular polygons with an increasing number of sides fitting it more and more closely." Courant goes on to give a recursion formula for the area of the inscribed regular polygon and defines the area of the circumscribed regular polygon based on that formula. He argues that π is the limit of both of these areas, and that Archimedes' method is closely related to our current concept of integral.

(5) The Compression Method (Baron, pages 37-38): "It is however, in the use which he makes of the compression method that Archimedes demonstrates the full power and elegance of the exhaustion techniques and it is here that their relation to the Cauchy-Riemann integral is most clearly exhibited. The method consists essentially in the establishment through some specialized geometric construction, of a monotonically ascending sequence, I_n, and a monotonically descending sequence, C_n, between which lies the magnitude S whose value is to be investigated. The terms I_n and C_n consist of the perimeters, surface areas or volumes of inscribed and circumscribed figures respectively and the relation $I_1 < I_2 < I_3 < \ldots < I_n < S < C_n < C_{n-1} < \ldots < C_2 < C_1$ is validated by a whole series of important convexity lemmas such as: (1) of lines which have the same extremities the straight line is the least; (2) lines (and surfaces) are concave in the same direction provided that all straight lines formed by joining any two points on the line (or surface) lie on the same side of the line (or surface); (3) if two lines (or surfaces) which have the same extremities be concave in the

same direction then that line (or surface) which is either wholly or partly enclosed within the other is the lesser of the two. By means of the particular construction adopted for I_n and C_n it is now shown that the difference $C_n - I_n$ can be made less than any assigned magnitude or that the ratio C_n/I_n can be made less than the ratio of the greater of any two assigned magnitudes to the lesser by a suitable choice of n. It is now necessary to determine some quantity K, such that $I_n < K < C_n$ for all values of n. We have thus, $I_n < S < C_n$ and $I_n < K < C_n$. Hence either $S = K$ or $S > K$ or $S < K$. The proof is completed by *reductio ad absurdum.* (i.e., assume $S > K$ and show that assumption that leads to a contradiction. Do the same with $S < K$. Since S is neither greater than nor less than K, it follows that $S = K$.)

(6) One wonders why Archimedes stopped at the 96-gon; certainly, his process could be iterated further. See the algebraic formulas below, which some think Archimedes used to compute his approximation of π (from Eves, p96):

(a) If s_k denotes the length of one side of a regular polygon of k sides inscribed in a circle of radius r, then $s_{2n} = \sqrt{2r^2 - r\sqrt{4r^2 - s_n^2}}$.

(b) If S_k denotes the length of one side of a regular polygon of k sides circumscribed about a circle of radius r, then $S_{2n} = \dfrac{2rS_n}{2r + \sqrt{4r^2 + S_n^2}}$.

(c) If p_k and P_k denote, respectively, the perimeters of the regular polygons of k sides inscribed in and circumscribed about the same circle, then $P_{2n} = \dfrac{2p_nP_n}{P_n + p_n}$, and $p_{2n} = \sqrt{p_nP_{2n}}$.

(7) Archimedes used two different approximations of $\sqrt{3}$ (one too large and one too small) as the basis for his upper and lower bounds for π. He gave no indication of where those fractions came from. Archimedes may have used the following approach for the approximations of roots. (Dijksterhuis, pages 229-238)

$$\sqrt{3} \text{ [add zero]} = 1 + (\sqrt{3} - 1) \text{ [write as a fraction]}$$

$$= 1 + \cfrac{1}{\cfrac{1}{\sqrt{3} - 1}} \text{ [rationalize the denominator of the new fraction]}$$

$$= 1 + \cfrac{1}{\cfrac{\sqrt{3} + 1}{2}} \text{ [remove integer part from denominator]}$$

$$= 1 + \cfrac{1}{1 + \cfrac{\sqrt{3} - 1}{2}} \text{ [write as a fraction]}$$

$$= 1 + \cfrac{1}{1 + \cfrac{1}{\left[\cfrac{2}{\sqrt{3} - 1}\right]}} \text{ [rationalize the denominator of the new fraction]}$$

$$= 1 + \cfrac{1}{1 + \cfrac{1}{\sqrt{3} + 1}} \text{ [add zero]}$$

$$= 1 + \cfrac{1}{1 + \cfrac{1}{2 + (\sqrt{3} - 1)}}$$

$$= \ldots\ldots,$$

but we've seen that $\sqrt{3} - 1$ before, on the first line. From the first and most recent lines,

$$\sqrt{3} - 1 = \cfrac{1}{1 + \cfrac{1}{2 + (\sqrt{3} - 1)}},$$

so we can replace that last $\sqrt{3} - 1$, giving

$$\sqrt{3} - 1 = \cfrac{1}{1 + \cfrac{1}{2 + \cfrac{1}{1 + \cfrac{1}{2 + (\sqrt{3} - 1)}}}},$$

again, giving

$$\sqrt{3} - 1 = \cfrac{1}{1 + \cfrac{1}{2 + \cfrac{1}{1 + \cfrac{1}{2 + \cfrac{1}{1 + \cfrac{1}{2 + (\sqrt{3} - 1)}}}}}},$$

and so on, for as long as we have the patience or need to do this. But there is a clear pattern taking place before eyes, namely, a repeated fraction with a repeating of 1's and 2's.
We can say

$$1 + (\sqrt{3} - 1) = \sqrt{3} = 1 + \cfrac{1}{1 + \cfrac{1}{2 + \cfrac{1}{1 + \cfrac{1}{2 + \cfrac{1}{1 + \cfrac{1}{2 + (\sqrt{3} - 1)}}}}}}.$$

We can make approximations of $\sqrt{3}$ by truncating the continued fraction; and we can make them as accurate as we wish by extending the continued fraction. So the first will be $\sqrt{3} \approx 1$, too small because we have left off the positive fraction. The second approximation is $\sqrt{3} \approx 1 + \frac{1}{1} = 2$, too large because we made the denominator of the fraction smaller, making the fraction itself larger.

The third is $\sqrt{3} \approx 1 + \cfrac{1}{1 + \frac{1}{2}} = \frac{5}{3}$, too small.

The fourth is $\sqrt{3} \approx 1 + \cfrac{1}{1 + \cfrac{1}{2 + \frac{1}{1}}} = \frac{7}{4}$, too large.

The fifth approximation is $\sqrt{3} \approx 1 + \cfrac{1}{1 + \cfrac{1}{2 + \cfrac{1}{1 + \frac{1}{2}}}} = \frac{19}{11}$, too small.

The sixth is $\sqrt{3} \approx 1 + \cfrac{1}{1 + \cfrac{1}{2 + \cfrac{1}{1 + \cfrac{1}{2 + \frac{1}{1}}}}} = \frac{26}{15}$, too large.

Continuing in this manner, we get the seventh: $\sqrt{3} \approx \frac{71}{41}$, small; the eighth: $\sqrt{3} \approx \frac{97}{56}$, large; the ninth: $\sqrt{3} \approx \frac{265}{153}$, small, and one of the approximations Archimedes used; the tenth: $\sqrt{3} \approx \frac{362}{209}$, large; the eleventh: $\sqrt{3} \approx \frac{989}{571}$, small; and the twelfth: $\sqrt{3} \approx \frac{1351}{780}$, large and the other approximation that Archimedes used. The same algorithm can be used to approximate $\sqrt{2}$ and roots of other positive integers.

(8) Eudoxus developed the theory of proportions used by Archimedes and his contemporaries. This theory is considered by some to be equivalent to modern theory of real numbers (see Stillwell, p 39 and Kline, pp 68-73.); at the very least, it is interesting to see how the theory of proportions allowed the Greeks to get past the difficulties caused by the discovery of incommensurable magnitudes (basically, irrational numbers are not the products of rational numbers).

Special implementation suggestions: See the implementation suggestions in the introduction to Historical Projects. For this project, you may want to consider giving Archimedes' work as a reading assignment for the whole class with lecture/discussion the following day. You could then ask all the students to complete the extension of his approximation. This would be a fairly efficient way of familiarizing students with one of the classic works and persons of mathematics.

Title: Zeno's Paradoxes

Authors: Charles Jones, Grinnell College
Mic Jackson and Will Carter, Earlham College

Problem Statement: Calculus is not merely a technical tool: it is "a collection of abstract mathematical ideas which have accumulated over long periods of time." Although there are many themes in the development of calculus, they all contain the common element of "the conflict between the demands of mathematical rigor imposed by deductive logic and the essential nature of the infinitely great and the infinitely small perpetually leading to paradox and anomaly." (Baron, pages 1-3)

Zeno of Elea described a number of paradoxes, including *The Dichotomy*, *Achilles and the Tortoise*, *The Arrow*, and *The Stadium*, which seem to prove the impossibility of motion by using an impeccable chain of logic. In each paradox, Zeno made a tacit assumption either that time and space could be infinitely divided or that they were made up of indivisible elements. (Many Greeks called these indivisible elements atoms; today we tend to call them infinitesimals.) He then showed that motion was impossible under either assumption. "Neither the infinite divisibility of the straight line nor the line as an infinite set of discrete points seemed to permit rational conclusions about motion." (Kline, page 992) At any rate, these paradoxes proved sufficient to frighten mathematicians away from using the idea of "infinity" until the seventeenth century. (Hooper, page 235)

Mathematicians have developed rigorous answers to these paradoxes only in relatively recent times. It should be noted that none of the answers are satisfactory to all mathematicians. The most commonly accepted answers have to do with infinite series and the definition of the derivative as an instantaneous rate of change.

Your assignment:

1. Search in your science library to find a statement of *The Dichotomy* and of *The Arrow*. You may want to start with some of the references cited above. Another good reference is Cajori. Your instructor has complete citations for each reference. For each paradox, try to determine whether Zeno made the assumption of infinite divisibility or the assumption of the existence of indivisible elements. Write a modern interpretation of each paradox: convince a rational person who had never studied the derivative or infinite series to agree that motion is in fact only an illusion, regardless of which initial assumption is made.

2. Complete the following exercises.

a. (1) What is $\left[\frac{1}{10}\right]^m$, for $m = 1, 2, 3, 10$? What if $m \to \infty$?
(Hint: Consider larger values of m.)
(2) Repeat (a1) for $(0.5)^m$.
(3) Repeat (a1) for $(0.9)^m$.
(4) Repeat (a1) for $(1.1)^m$.

b. (1) Simplify $(1 - x)(1 + x + x^2 + \ldots + x^n)$.

(Hint: Try $(1-x)(1)$, $(1-x)(1+x)$, $(1-x)(1+x+x^2)$, $(1-x)(1+x+x^2+x^3)$, etc., to see the pattern.)

(2) What answer did you get ? How can you use this to represent the sum $1 + x + x^2 + \ldots + x^n$ in a manageable way?

c. Using your answers to exercises (a) and (b2), evaluate $(1 + x + x^2 + \ldots + x^n)$ where:

(1) $x = \frac{1}{10}$, and $n = 9$, (2) $x = \frac{1}{2}$, and $n = 9$, (3) $x = 0.9$, and $n = 9$,

(4) $x = 1.1$, and $n = 9$, (5) $x = \frac{1}{10}$, and $n \to \infty$, (6) $x = \frac{1}{2}$, and $n \to \infty$,

(7) $x = 0.9$, and $n \to \infty$, (8) $x = 1.1$, and $n \to \infty$.

3. The argument of the Dichotomy implies that a runner in a 100 meter race never crosses the finish line. Zeno's reasoning was that the runner must reach the first halfway point (50 meter mark). From there, our runner must reach the second halfway-to-the-finish point (75 meter mark). Then, our runner must reach the third halfway-to-the-finish point. This continues indefinitely; there are an infinite number of these halfway-to-the-finish points, and there is a positive distance between any two of them. Since the distance between consecutive halfway-to-the-finish points is positive, the time required to go from one point to the other is also positive. So, our runner never reaches the finish line since the time required is the sum of infinitely many positive quantities. This interpretation of the Dichotomy is used in the following problems; assume the runner travels at a constant speed of 10 m/sec.

a. Using information from exercise 2c, explain why the runner can reach the finish line. When does the runner cross the finish line?

b. How far from the finish line is the (1) 10th halfway-to-the-finish point ?
(2) the 20th ?
(3) the 100th ?

c. Which of the halfway-to-the-finish points occurs during the final $\left(\frac{1}{1000}\right)^{\text{th}}$ second of the race?

4. The following exercises are optional.

a. Explain Achilles and the Tortoise, assuming that Achilles runs 10 times as fast as the tortoise, and the tortoise has a 900 meter head start.

b. Come up with questions analogous to exercises 3a through 3c for this paradox, and answer them using what you learned in exercises 2a through 2c.

Your report will be evaluated for mathematical correctness, thoroughness and clarity, as well as for the normal criteria of any written submission. Be sure to cite any library resources you use.

Information for the instructor only:

Project abstract: This project is designed to introduce students to infinite series and some early history of how humans dealt with the notion of infinity. It is one that naturally encourages young students to discuss and debate their perceptions of the world. We recommend assigning this project before dealing with series in class; the only nonalgebraic concept they need is an understanding of the notation $m \to \infty$. The purposes of the project are to allow students to experiment with the key idea of series (that the sum of a series is the limit of the sequence of partial sums) before they are given the formal definitions, to see series in the interesting context of Zeno's paradoxes, and to consider how series and the definition of the derivative as an instantaneous rate of change provide answers to Zeno's paradoxes.

Newton, Leibniz, Euler and other mathematicians involved in the early development and exploitation of the calculus all approached their work with the assumption of the existence of infinitesimals, basically ignoring the philosophical problems pointed out by Zeno and others. As paradoxes and inconsistencies were encountered in the exploitation of the derivative and integral, Cauchy, Weierstrass, and others argued for a theoretical foundation upon which a consistent theory underlying calculus could be developed. The axiomatization of the real number system, and hence the rigorous theory which underlies calculus, is based on the modern notion of limit which is related to the assumption of infinite divisibility. Until quite recently infinitesimals were dead as a theory but an American mathematician, Abraham Robinson, articulated a consistent calculus based on the assumption of infinitesimals around 1960. (See Henle and Kleinberg or Hurd and Loeb.)

Because it is difficult to find clear descriptions of the four paradoxes, I've included descriptions below from different sources. You may want to give some form of these descriptions to your students and ask them to try to rewrite the paradoxes for a modern audience as part 1 of their assignment.

Dichotomy
Cajori, page 2 - You cannot traverse an infinite number of points in a finite time. You must traverse the half of any given distance before you traverse the whole, and the half of that again before you can traverse the whole, and the half of that again before you can traverse it. This goes on *ad infinitum*, so that (if space is made up of points) there are an infinite number in any given space, and it cannot be traversed in a finite time.
Eves, page 288 - If a straight line segment is infinitely divisible, then motion is impossible, for in order to traverse the line segment it is necessary first to reach the midpoint, and to do this one must first reach the one-quarter point, and to do this one must first reach the one-eighth point, and so on, *ad infinitum*. It follows that motion can never even begin (if space is infinitely divisible).

Achilles and the Tortoise
Cajori, page 2 - Achilles must first reach the place from which the tortoise started. By that time the tortoise will have moved on a little way. Achilles must then traverse that, and still the tortoise will be ahead. Achilles is always nearer, but never catches up to the tortoise (if space is infinitely divisible).

Arrow
Cajori, p2 - This third argument against the possibility of motion through a space made up

of points is that an arrow in any given moment (where moment is the smallest amount of time) of its flight must be at rest in some particular point.
Eves, page 288 - If time is made up of indivisible atomic instants, then a moving arrow is always at rest, for at any instant the arrow is in a fixed position. Since this is true every instant, it follows that the arrow never moves.

Stadium
Cajori, p3 - suppose there are three parallel rows of points lined up as shown in Figure 1.

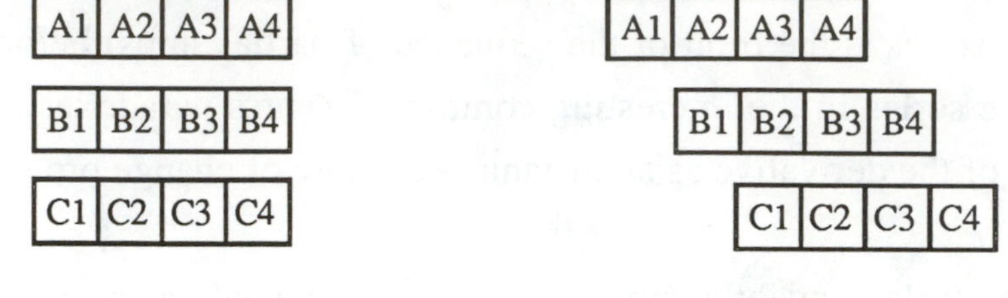

Figure 1 **Figure 2**

Of these row B is immovable, while rows A and C move in opposite directions with equal velocity so as to come into the positions represented in Figure 2. The movement of row C relative to row A will be double its movement relative to row B, or in other words, any given point in row C has passed twice as many points in row A as it has in row B. It cannot, therefore, be the case that an instant of time corresponds to the passage from one point to another.
Assume that there is some smallest instant of time. Begin with twelve bodies of equal size lined up as shown in Figure 1, then allow the described movement for one instant. The fact that the movement of row A relative to row C is double that of row A to row B implies that the supposedly smallest instant of time could be halved.

Prerequisite skills and knowledge: Good algebraic skills and an understanding of the notation $m \to \infty$.

Essential/useful library resources: See the bibliography in the introduction to Historical Projects.

Essential/useful computational resources: None are required, although a scientific calculator might save the students some time.

Example of acceptable solutions: For problem (1) students should give for each of the paradoxes: an acceptable statement, assuming that you have not furnished one; a description of which assumption was made; an accurate interpretation of the situation using clear narrative and explanatory diagrams where helpful; and an explanation why the described situation under the given assumption leads to the conclusion that motion is impossible.

For problems 2 and 3, we've included sample solutions:

(2a1) $\left[\frac{1}{10}\right]^1 = 0.1$, $\left[\frac{1}{10}\right]^2 = 0.01$, $\left[\frac{1}{10}\right]^3 = 0.001$, and $\left[\frac{1}{10}\right]^{10} = 0.0000000001$.

It appears that as m gets arbitrarily large, $\left[\frac{1}{10}\right]^m$ gets arbitrarily small,

or as $m \to \infty$, $\left[\frac{1}{10}\right]^m \to 0$.

(2a2) $\left[\frac{1}{2}\right]^1 = 0.5,\quad \left[\frac{1}{2}\right]^2 = 0.25,\quad \left[\frac{1}{2}\right]^3 = 0.125,\quad \left[\frac{1}{2}\right]^{10} = \frac{1}{1024} \approx 0.000976562.$

It appears that as m gets arbitrarily large, $\left[\frac{1}{2}\right]^m$ gets arbitrarily small,

or as $m \to \infty$, $\left[\frac{1}{2}\right]^m \to 0$.

(2a3) $(0.9)^1 = 0.9,\ (0.9)^2 = 0.81,\ (0.9)^3 = 0.729,\ (0.9)^{10} \approx 0.34868.$
It appears that as m gets arbitrarily large, $(0.9)^m$ gets arbitrarily small,
or as $m \to \infty$, $(0.9)^m \to 0$.

(2a4) $(1.1)^1 = 1.1,\ (1.1)^2 = 1.21,\ (1.1)^3 = 1.331,\ (1.1)^{10} \approx 2.6.$
It appears that as m gets arbitrarily large, $(1.1)^m$ gets arbitrarily large,
or as $m \to \infty$, $(1.1)^m \to \infty$.

(2b1) $(1-x)(1) = (1-x),$
$(1-x)(1+x) = 1-x^2,$
$(1-x)(1+x+x^2) = 1-x^3,$
$(1-x)(1+x+x^2+x^3) = 1-x^4.$
The pattern is $(1-x)(1+x+\ldots+x^n) = 1-x^{(n+1)}$.

(2b2) This is the key step:
$(1-x)\,(1+x+x^2+\ldots+x^n) = 1-x^{(n+1)}$ implies $1+x+x^2+\ldots+x^n = \frac{1-x^{(n+1)}}{1-x}$.

(2c) Using the result of b2, we can quickly evaluate each sum for the given value of x and n.

(2c1) $\frac{1-(0.1)^{(10)}}{1-(0.1)} = \frac{0.9999999999}{0.9} = 1.111111111$

(2c2) $\frac{1-(0.5)^{(10)}}{1-(0.5)} = \frac{0.9990234375}{0.5} = 1.998046875$

(2c3) $\frac{1-(1.9)^{(10)}}{1-(0.9)} = \frac{0.6513215599}{0.1} = 6.513215599$

(2c4) $\frac{1-(1.1)^{(10)}}{1-(1.1)} = \frac{-1.59374246}{-0.1} = 15.9374246$

(2c5) As $n \to \infty$, $\frac{1-(0.1)^{(n+1)}}{1-(0.1)} \to \frac{1-0}{0.9} = \frac{10}{9} = 1.111\ldots$

(2c6) As $n \to \infty$, $\frac{1-(0.5)^{(n+1)}}{1-(0.5)} \to \frac{1-0}{0.5} = 2$

(2c7) As $n \to \infty$, $\frac{1-(0.9)^{(n+1)}}{1-(0.9)} \to \frac{1-0}{0.1} = 10$

(2c8) As $n \to \infty$, $\frac{1-(1.1)^{(n+1)}}{1-(1.1)} \to \frac{-\infty}{-0.1} = \infty$

(3a) Exercises 2c5-7 show that the sum of an infinite sequence of numbers which become arbitrarily small can equal a finite number. Our runner travels half the remaining distance in each instant: hence, the distance she travels is $(0.5) + (0.5)^2 + (0.5)^3 + (0.5)^4 + \cdots + (0.5)^n$ which approaches $2 - 1 = 1$ (= the length of the race) as n increases without bound.

(3b) Since the distance remaining to the finish line is 100 meters minus the distance traveled. So the distance remaining is 100m times $\{1 - [(0.5) + (0.5)^2 + (0.5)^3 + (0.5)^4 + \cdots + (0.5)^n]\}$ = 100m times

$\{1 - [\frac{1 - (0.5)^{(n+1)}}{1 - (0.5)} - 1]\}$, where n is the number of the current halfway-to-the-finish-point.

(3b1) 100m times $(1 - [\frac{1 - (0.5)^{(11)}}{1 - (0.5)} - 1]) \approx 0.09765624$ meters.

An easier way of doing this is $\left[\frac{1}{2}\right]^{10} \times 100 \approx 0.0976562$ meters.

(3b2) $\left[\frac{1}{2}\right]^{20} \times 100 \approx 0.0000953$ meters

(3b3) $\left[\frac{1}{2}\right]^{100} \times 100 \approx 7.9 \times 10^{-29}$ meters

(3c) At 10 meters per second our runner could cover 0.01 meters, or 1 centimeter, during the final $\left(\frac{1}{1000}\right)^{\text{th}}$ second of the race. So we need to find an integer n such that $\left[\frac{1}{2}\right]^{n} \times 100 \leq 0.01$, or $2^n \geq 10000$: a little experimentation shows that this inequality is true if $n \geq 14$. So every halfway-to-the-finish point *after* the fourteenth occurs during the final $\left(\frac{1}{1000}\right)^{\text{th}}$ second of the race!

(4a) By the following reasoning, Achilles can never catch the tortoise . Call the tortoise's starting position point 1. By the time Achilles reaches point 1, the tortoise has moved to another position, call it point 2. By the time Achilles reaches point 2, the tortoise has moved to another position, call it point 3; this continues indefinitely. During the positive amount of time Achilles takes to move from point n to point $n+1$, the tortoise also moves some positive distance, to point $n+2$. Since there are an infinite number of these points, Achilles never catches the tortoise. Assume that the tortoise runs 10 meters per minute.

(4b) Some analogous questions (without answers).

a. Using information from exercise 2c, explain why the Achilles can and does catch the tortoise.

b. How far from the tortoise is Achilles when Achilles is at (1) point 10 ?
(2) point 20 ?
(3) point 100 ?

c. Which of Achilles' points occur during the final $\left(\frac{1}{1000}\right)^{\text{th}}$ minute of the race?

Conjectures we expect that some students will make: One of these assumptions, infinite divisibility or existence of indivisibles, must be correct, the other incorrect.

Questions for further exploration: Do some similar work with the Arrow or the Stadium.

References/bibliography/related topics: the other historical projects, convergence of infinite series

Special implementation suggestions: See the implementation suggestions in the introduction to Historical Projects.

Title: Archimedes' Determination of the Surface Area of a Sphere

Authors: Mic Jackson and Krista Briese, Earlham College

Problem Statement: In this particular proof, Archimedes used the assumption that certain processes could be carried on indefinitely (his way of dealing with the notion of infinity), Eudoxes' theory of proportions (his way of dealing with the problem of the existence of irrational numbers), and a form of proof known as *reductio ad absurdum.* One example of his use of the theory of proportions is his contention that, given two circles, it must be true that the ratio of the circumference to the diameter of one is equal to the ratio of the circumference to the diameter of the other: $\frac{C_a}{d_a} = \frac{C_b}{d_b}$. We would make the equivalent statement that the circumference of any circle is equal to the diameter of the circle multiplied by the number π; $C = \pi d$, or $\frac{C}{d} = \pi$ for any circle. Hence, anyone attempting to follow Archimedes' work will also need to be competent with the rules and concepts of proportion which where characteristic of ancient Greek mathematics. Some are listed below:

(1) If $\frac{a}{b} = \frac{b}{c}$, then $\frac{a}{c} = \frac{a^2}{b^2}$. **(2)** If $\frac{a}{b} = \frac{b}{c} = \frac{c}{d}$, then $\frac{a}{d} = \frac{a^3}{b^3}$.

(3) If $\frac{a}{b} = \frac{m}{n}$ and $\frac{b}{c} = \frac{p}{q}$, then $\frac{a}{c} = \frac{mp}{nq}$.

(4) If $\frac{a}{b} = \frac{c}{d}$, then (a) $\frac{a}{c} = \frac{b}{d}$, (b) $\frac{b}{a} = \frac{d}{c}$, (c) $\frac{a+b}{b} = \frac{c+d}{d}$,

(d) $\frac{a-b}{b} = \frac{c-d}{d}$ if $a > b$, (e) $\frac{a}{a-b} = \frac{c}{c-d}$ if $a > b$,

(f) $\frac{a+b}{a} = \frac{c+d}{c}$, (g) $\frac{b}{a-b} = \frac{d}{c-d}$, (h) $\frac{a-b}{a} = \frac{c-d}{c}$.

(5) If $\frac{a}{b} > \frac{c}{d}$, then (a) $\frac{a}{c} > \frac{b}{d}$, (b) $\frac{b}{a} < \frac{d}{c}$, (c) $\frac{a+b}{b} > \frac{c+d}{d}$,

(d) $\frac{a-b}{b} > \frac{c-d}{d}$, (e) $\frac{a}{a-b} < \frac{c}{c-d}$, (f) $\frac{a+b}{a} < \frac{c+d}{c}$,

(g) $\frac{b}{a-b} < \frac{d}{c-d}$, (h) $\frac{a-b}{a} > \frac{c-d}{c}$; (i) if also $c > d$, then $a > b$.

(6) If $a > b$, and c is "any magnitude homogeneous with a and b," then $\frac{a}{b} > \frac{a+c}{b+c}$.

Your task is to justify the 13 statements highlighted in **bold** face in Archimedes' proof that the surface area of a sphere is equal to $4\pi r^2$. To justify in this case means to provide the details that Archimedes left out so that the proof is more easily readable. You need to show how each statement follows from known definitions, axioms and/or theorems and from previous statements in this proof. You will need to make geometrical arguments and use some of your knowledge of proportions and infinite series to understand Archimedes. In each case, be sure you explain what Archimedes meant and why he felt the phrase was necessary.

Archimedes' determination of the surface area of a sphere.

(Adapted from *The Origins of the Infinitesimal Calculus*, by Margaret Baron, pages 38-41.[1])

PART I

In trying to determine the surface of a sphere Archimedes' basic plan was to find a circle whose area was identical to the surface area of a given sphere. He began with a maximal circular cross-section of the sphere and inscribed within that circle, which we will call circle S, a regular polygon of n sides (Figure A). He then rotated the circle S and regular n-gon about a diameter (AC in the figure). He considered the solid resulting from the rotation of the n-gon to be made up of successive *conical frusta* ($BXYZ$ is a cross section of such a frustrum.), and approximated the surface area of the sphere with the sum of the surface areas of all the conical frustra.

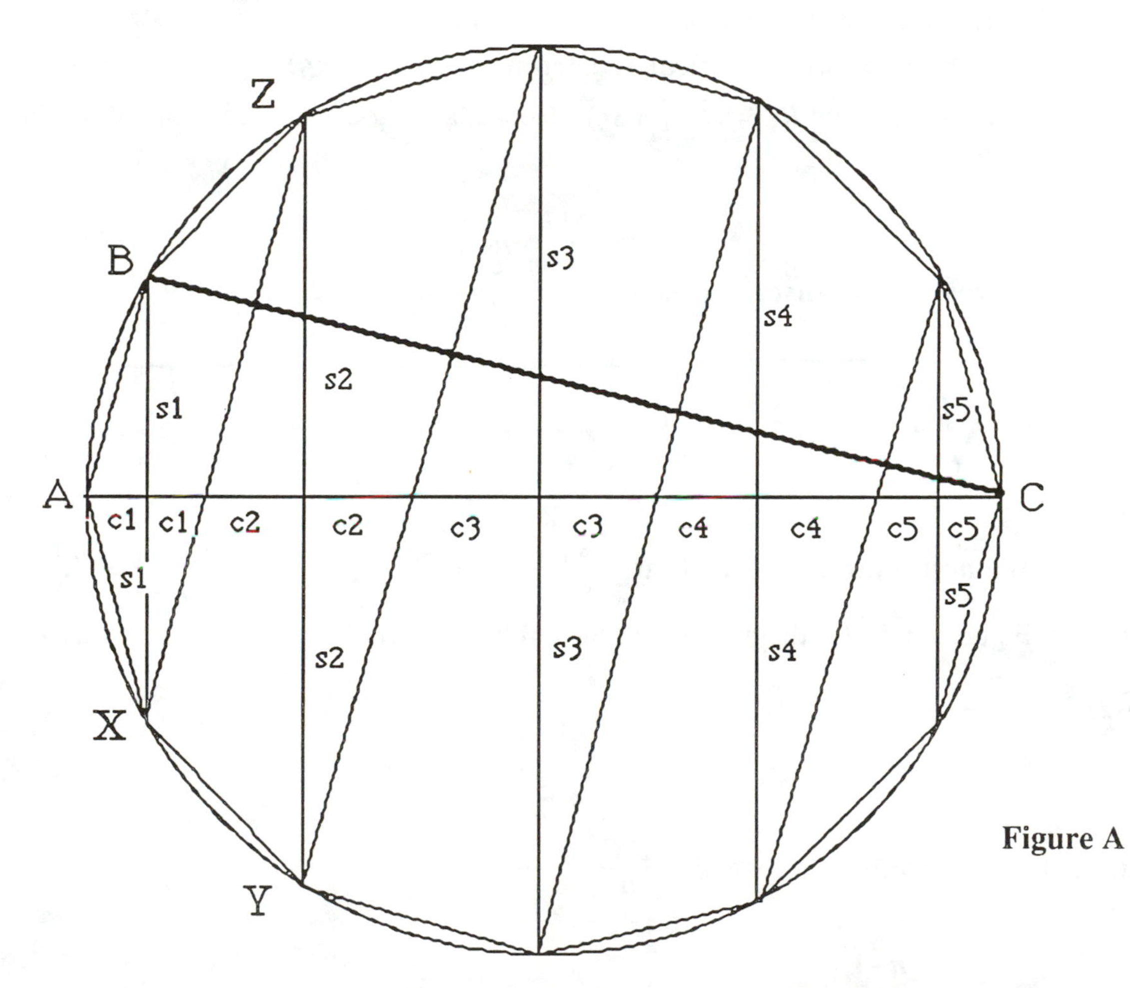

Figure A

He next determined the area of each conical frustra by constructing for each a circle of equal area (Figure B). Finally, he replaced the sum of all of these circular areas with the area of a single circle.

---

[1] Adapted with the permission of Macmillan Publishing Company, a Division of Macmillan, Inc. from THE ORIGINS OF THE INFINITESIMAL CALCULUS by Margaret Baron. Copyright ©1969 Pergamon Press.

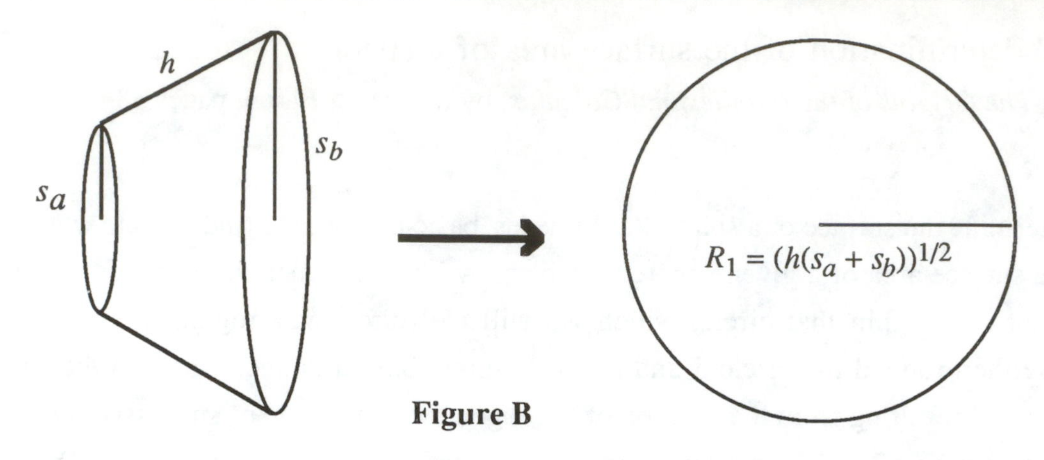

Figure B

To see how Archimedes replaced each conical frustrum with a circle of equal area, consider that if the radii of the frustrum are s_a and s_b and the slant height is h then

(1) the surface area of a single conical frustrum is $\pi h(s_a + s_b)$.

Thus $(s_a + s_b)h = (R_1)^2$, and $R_1 = \sqrt{h(s_a + s_b)}$. Hence, adding all surface areas of all the conical frustra,

(2) the surface area of the inscribed solid is $\pi \sum_{r=1}^{n-1} R_r^2 = \pi R^2$,

where $R = \sqrt{\sum_{r=1}^{n-1} R_r^2} = \sqrt{\sum_{r=1}^{n-1} h(s_r + s_{r+1})} = \sqrt{2h \sum_{r=1}^{n-1} s_r}$.

(3) It follows from similar triangles that $\frac{s1}{c1} = \frac{s2}{c2} = \frac{s3}{c3} = \frac{s4}{c4} = \cdots = \frac{s_{n-1}}{c_{n-1}} = \frac{BC}{h}$.

Now (from Figure A), if d is the diameter of the circle in which the regular n-gon is inscribed, then $d = 2\sum_{r=1}^{n-1} c_r$.

(4) Using rules of proportion, we get $\frac{BC}{h} = \frac{2\sum_{r=1}^{n-1} s_r}{d}$.

(5) It follows that $2h \sum_{r=1}^{n-1} s_r = d \cdot BC$, **and** $R = \sqrt{d \cdot BC}$ (the radius of a circle whose area is identical to the surface area of the inscribed solid). Since $BC < d$ it follows that $R < d$, and

(6) the surface area of the inscribed solid is less than that of the circle with radius d.

It is useful to recall that d is the diameter of circle S in which we inscribed the regular n-gon. We are now considering a circle whose area is 4 times greater than that of the original circle S.

PART II

Having produced a lower bound for the surface area of a sphere in Part I, Archimedes determined here to produce an upper bound. He circumscribed a regular n-gon about the circle S and then considered this n-gon as inscribed in a circle S' with diameter d' ($d' > d$).

(7) The surface area of the solid circumscribed about circle S can be written down immediately as equal to the area of a circle with radius $R' = \sqrt{d' \cdot B'C'}$.

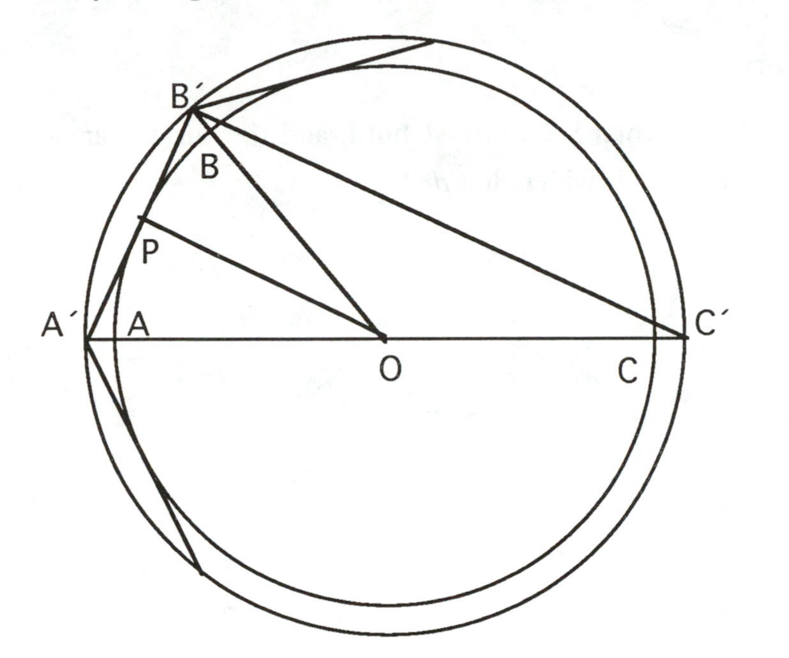

Figure C

(8) In triangle $A'B'C'$, $A'P = PB'$ and $A'O = OC'$, hence $B'C' = 2PO = d$. Finally, since $B'C' = d$,

(9) the surface area of the circumscribed solid is greater than that of the circle with radius d.

PART III

Archimedes' last step in many of his demonstrations is a proof by *reductio ad absurdum* ; what follows is fairly typical. Let S_n and S'_n denote the surface areas of the inscribed and circumscribed solids (formed by the rotation of a regular n-gon) respectively, and h_n and h'_n denote the slant height of each conical frustrum. Finally, let S denote the surface area of the sphere and K the area of a circle with radius d.

(10) It follows that $S_n < S < S'_n$ and $S_n < K < S'_n$.

(A) Assume that $S < K$. Let x and y be such that $\frac{x}{y} < \frac{K}{S}$.

(11) There exists a counting number n be such that $\frac{S'_n}{S_n} = \frac{{h'_n}^2}{{h_n}^2} < \frac{x}{y}$.

Hence, $\frac{S'_n}{S_n} < \frac{K}{S}$, which is impossible since $S'_n > K$ and $S_n < S$.

(12) So it must not be true that S is less than K.

(B) Assume that $S > K$. Let x and y be such that $\frac{x}{y} < \frac{S}{K}$. Let n be such that $\frac{S'_n}{S_n} = \frac{h'_n{}^2}{h_n{}^2} < \frac{x}{y}$. Hence, $\frac{S'_n}{S_n} < \frac{S}{K}$, which is impossible since $S'_n > S$ and $S_n < K$. So it is not possible for S to be greater than K.

(C) **(13) Since $S \nless K$ and $S \ngtr K$, then $S = K$ must hold,** and the surface area of a sphere of diameter d is equal to the area of a circle with radius d.

Information for the instructor only:

Problem abstract: This problem provides a good example of Archimedes' use of geometry and proportions and his assumption that he could continue a process "as long as necessary." In this case the process involved circumscribing and inscribing regular n-gons about a circle. He employed the "compression method" in which he demonstrated that the desired value lay between a lesser monotonically increasing sequence and a greater monotonically decreasing sequence. (See the introduction to Historical Projects.) Without addressing the actual values of the sequences and tacitly assuming the convergence of both as obvious, he made good use of Eudoxus' theory of proportions to establish the desired result. As usual, he made no mention of π, but stated his result in terms of its relationship to known values.

Prerequisite skills and knowledge: high school algebra (including theory of proportions), high school geometry, convergent sequences, infinite series.

Essential/useful library resources: See the bibliography in the introduction to Historical Projects.

Essential/useful computational resources: none

Example of an acceptable approach:

(1) the surface area of a single conical frustrum is $\pi h(s_a + s_b)$

The surface area of a right circular cone with base radius R and slant height L is πRL. The surface area of a frustrum of that cone is the surface area of the cone less the surface area of the part cut away.

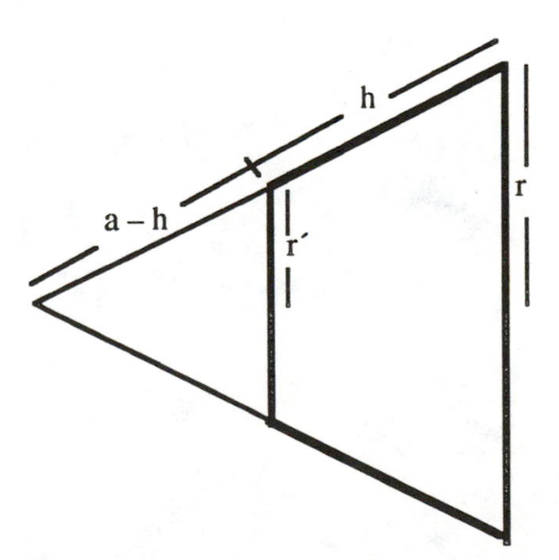

Figure D

Considering the conical frustrum formed by the rotation about its centerline of the emboldened trapezoid in Figure D, we find

$$Af = \pi\, ar - \pi\,(a-h)r' = \pi\, a(r-r') + \pi\, hr' = \text{(by similar triangles) } \pi\frac{hr}{r-r'}(r-r') + \pi\, hr' = \pi\, h(r+r').$$

(2) the surface area of the inscribed solid is $\pi \sum_{r=1}^{n-1} R_r^2 = \pi R^2$,

where $R = \sqrt{\sum_{r=1}^{n-1} R_r^2} = \sqrt{\sum_{r=1}^{n-1} h(s_r + s_{r+1})} = \sqrt{2h \sum_{r=1}^{n-1} s_r}$.

For the 12-gon in Figure A, there are $n = 6$ conical frustra; the sum of their surface areas will give the surface area of the inscribed solid. Further, the pattern for this particular example will hold for any value of n, so it will be easy to move to a general expression. From the above result for the surface area of a frustrum it follows that the surface area for the particular example is

$$A = \pi h(s_0+s_1) + \pi h(s_1+s_2) + \pi h(s_2+s_3) + \pi h(s_3+s_4) + \pi h(s_4+s_5) + \pi h(s_5+s_6)$$

$$= \pi h(0+s_1) + \pi h(s_1+s_2) + \pi h(s_2+s_3) + \pi h(s_3+s_4) + \pi h(s_4+s_5) + \pi h(s_5+0) = 2\pi h \sum_{r=1}^{5} s_r.$$

(3) It follows from similar triangles that $\frac{s_1}{c_1} = \frac{s_2}{c_2} = \frac{s_3}{c_3} = \frac{s_4}{c_4} = \cdots = \frac{s_{n-1}}{c_{n-1}} = \frac{BC}{h}$.

Again, if we consider the particular example in Figure A, the result is easily generalizable to any n-gon. Establishing the particular results is messy enough.

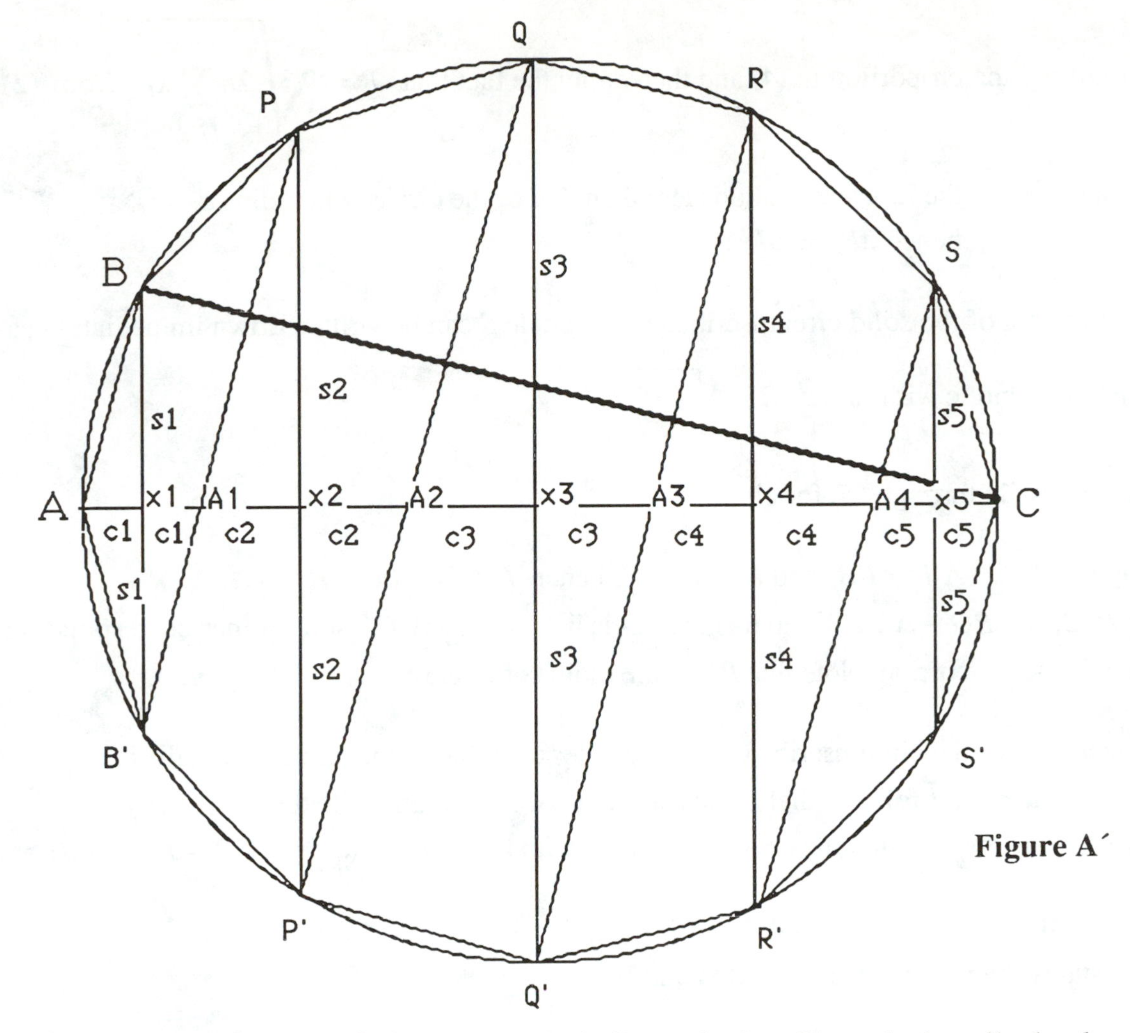

Figure A´

First, we need to establish which triangles are similar and why. Since the inscribed polygon is regular it follows that $\angle B'PP'$ is congruent to $\angle P'QQ'$, which implies that segment PB' is parallel to segment $P'Q$, and that $\angle PA_1X_2 \cong \angle QA_2X_3$, and finally that $\Delta PA_1X_2 \sim \Delta QA_2X_3$. The same reasoning can be applied to show: $\Delta BAX_1 \sim \Delta PA_1X_2 \sim \Delta QA_2X_3 \sim \Delta RA_3X_4 \sim \Delta SA_4X_5$. Further, $\angle ACB \cong \angle ABB'$ because they are inscribed angles cutting off equal arcs, so $\Delta CAB \sim \Delta BAX_1$, and to all the other triangles listed above. The list of equal ratios immediately follows.

(4) Using rules of proportion, we get $\dfrac{BC}{h} = \dfrac{2\sum_{r=1}^{n-1} s_r}{d}$.

If $\frac{a}{b} = \frac{c}{d} = \frac{e}{f}$, then $\frac{a+c}{b+d} = \frac{a}{b} = \frac{c}{d} = \frac{e}{f}$. This generalizes to any list of equal ratios.

(5) It follows that $2h\sum_{r=1}^{n-1} s_r = d \cdot BC$, and $R = \sqrt{d \cdot BC}$.

Cross-multiply the proportion in (4) and then apply the fact that $R = \sqrt{2h \sum_{r=1}^{n-1} s_r}$ from (2).

(6) The surface area of the inscribed solid is less than that of the circle with radius d.
$R^2 = d \cdot BC < d^2$; hence $\pi R^2 < \pi d^2$.

(7) The surface area of the solid circumscribed about circle S can be written down immediately as equal to the area of a circle with radius $R' = \sqrt{d' \cdot B'C'}$.

Follow arguments (1) through (5) for circle O'.

(8) In triangle $A'B'C'$, $A'P = PB'$ and $A'O = OC'$, hence $B'C' = 2PO = d$.
In Figure C, $\Delta A'PO \sim \Delta A'B'C'$ and side $A'O$ is half of side $A'C'$, so all other corresponding sides are in the same proportion. Note that PO is the radius of circle S.

(9) the surface area of the circumscribed solid is greater than that of the circle with radius d.
Since $B'C' = d < d'$, it follows that the surface area of the circumscribed solid [$\pi (R')^2 = \pi (d')(B'C')$] is greater than the area of the circle with radius d. [$\pi(d')(B'C') = \pi (d')(d) > \pi \cdot d^2$]

(10) It follows that $S_n < S < S'_n$ and $S_n < K < S'_n$.
This restates the results of from Parts I and II.

(11) There exists a counting number n such that $\frac{S'_n}{S_n} = \frac{h'_n{}^2}{h_n^2} < \frac{x}{y}$.
It is reasonable that $h_n' > h_n$ and $(h_n')^2 > h_n^2$ for all $n > 2$. The intuitively obvious but actually tricky argument here is that the positive difference $h_n' - h_n$ gets arbitrarily close to 0 as n increases and that $\frac{h'_n{}^2}{h_n^2}$ gets arbitrarily close to 1. I guess that is what Archimedes basically reasoned. So if $\frac{x}{y}$ is a fixed value greater than 1 it follows that if we continued Archimedes' process, we would eventually construct regular inscribed and circumscribed n-gons such that $\frac{h'_n{}^2}{h_n^2} < \frac{x}{y}$.

(12) So it must not be true that S is less than K.
The assumption that $S < K$ led to a false conclusion.

(13) Since $S \not< K$ and $S \not> K$, then $S = K$ must hold.
This follows from the trichotomy property which says that if we have two magnitudes S and K, then exactly one of the following statements is true: $S = K$, $S < K$, or $S > K$.

Conjectures we expect that some students will make: Archimedes did not talk about π, but ratios of magnitudes, why?

Questions for further exploration: Archimedes made a number of other significant determinations, most of which can be found in the library resources listed. A bright student would gain from tackling one on her own. Most students could gain from reading about Archimedes and his accomplishments.

A calculus founded on the notion of infinitesimals, rather than currently accepted limit theory, known as "non-standard analysis," (see Henle and Kleinberg) could provide the unusually bright and motivated student with some food for thought.

References/bibliography/related topics: A lengthy series of articles entitled "The History of Zeno's Arguments on Motion," by Florian Cajori, the *American Mathematical Monthly* published during the year of 1915 gives thorough coverage of the philosophical problems related to the very small and the very large.

Eudoxus developed the theory of proportion used by Archimedes and his contemporaries. This theory is considered by some to be equivalent to modern theory of real numbers (see Stillwell, page 39); at the very least, it is interesting to see how the theory of proportions allowed the Greeks to get past the difficulties caused by the discovery of incommensurable magnitudes which basically showed the existence of numbers (called irrational) that are not the products of rational numbers.

Special implementation suggestions: See the implementation suggestions in the introduction to Historical Projects.

Title: Newton's Investigation of Cubic Curves

Author: Jeffrey Nunemacher, Ohio Wesleyan University

Problem Statement: This project explores a portion of the classification of cubic curves, which was carried out by Isaac Newton in the late seventeenth century. The general cubic curve in two variables is defined by the equation

$$ax^3 + bx^2y + cxy^2 + dy^3 + ex^2 + ixy + jy^2 + mx + ny + p = 0 \qquad (1)$$

where a to p are real constants. We assume that at least one of the initial coefficients a to d is nonzero so that the curve is a legitimate cubic. There are many more varieties of cubics than of quadratics ($ax^2 + bxy + cy^2 + dx + ey + f = 0$) which turn out to be conic sections usually, although the degenerate cases of single lines, two parallel or intersecting lines, single points, and the null set can also occur. Newton classified nondegenerate cubics into 72 different species according to their asymptotic behavior. Actually, according to his scheme there should be 78 species; Newton missed six which were found by later authors. Although it is too complicated to look carefully at his complete classification, we can learn quite a bit by using calculus and algebra to redo a portion of his investigation. Proceed as follows:

1) To get an idea of the range and beauty of cubic curves, draw graphs of the famous curves listed below which have occurred in particular geometric or algebraic settings (and thus acquired names). You will find it useful to use a computer algebra package which can solve such equations for y in terms of x and then produce a graph of the result. However, be careful with the folium; you will need to do a parametric plot to get at this one properly. Be sure that your graph shows all features of the curves; you may need to zoom outwards to see the global picture.

a) $y(1 + x^2) = 1$, the witch of Maria Agnesi;
b) $y^2(2 - x) = x^3$, the cissoid of Diocles;
c) $x^3 + y^3 = 1$, the Fermat curve for $n = 3$;
d) $x^3 + y^3 - 3xy = 0$, the folium of Descartes.

2) Consider the special case $y = f(x) = ax^3 + bx^2 + cx + d$, where $a \neq 0$, b, c, and d are fixed real numbers. Determine the possible shapes for the graph in terms of appropriate functions of coefficients. It will be helpful to consider the solutions of $f'(x) = 0$. Sketch the various possibilities.

3) In the same way explore the shapes of the curves $xy = f(x) = ax^3 + bx^2 + cx + d$ for $a \neq 0$. Again determine the shapes in terms of the coefficients. You will need to decide when the equation $f(x) = 0$ has three positive or three negative solutions.

4) Continue your look at special cases by considering curves of the form $y^2 = f(x) = ax^3 + bx^2 + cx + d$ for $a \neq 0$. This time you should find five quite different shapes depending on the nature of the

roots of $f(x)$. Two of them look very similar but there is a crucial difference (what?). Find simple examples of each type, e.g., $y^2 = x^3 - x$ for the case of three distinct real roots. Sketch a graph of each shape.

5) Newton was able to reduce the general cubic equation (1) to one of four simpler types of equations by a suitable change of coordinates. These four types are:

a) $y = f(x) = ax^3 + bx^2 + cx + d$;
b) $xy = f(x) = ax^3 + bx^2 + cx + d$;
c) $y^2 = f(x) = ax^3 + bx^2 + cx + d$;
d) $xy^2 + ey = f(x) = ax^3 + bx^2 + cx + d$.

As you see, the first three types are those which we studied in parts 2) - 4). The first two types each contribute one species to the 78 species of nondegenerate cubic curve, the third type five species, and the fourth type the remaining 71 species.

Given the results of parts 2) and 3) above, you may be surprised that Newton counted these curves as contributing only one species each to his catalog. The reason lies in the fact that "bumps" can be smoothed out by a linear change of coordinates, which Newton allowed, so they cannot be distinguished within his classification scheme. As an example, start with the curve $y = x^3 - x$ and perform the change of variables $x' = x, y' = y + x$. Sketch the curve in the original x, y coordinates and in the new x', y' coordinates. What do you observe?

6) We have studied the first three types of cubic curves in parts 2-4. The fourth type is quite complicated in general, but notice that it is possible to solve for y in terms of x. Newton did this by multiplying through by x and then completing the square. Make any general statements possible about this fourth type of cubic. For example, how many points on the curve can lie above each x value? How many intervals on the x-axis can have no points above them? How many asymptotes are possible?

Another way for the curve to be degenerate (aside from having no cubic terms) is if the defining equation factors nontrivially into a product of terms, e.g., $xy^2 + y = x^3 - x$, which factors as $(x + y)(1 + xy - x^2) = 0$. Give a complete list of the types of cubic curves which are degenerate in this way.

Finally, use a computer to explore the nature of the type d) cubic. Try to find examples which have 1, 2, 3, and 4 branches. Can you find ten substantially different examples?

Information for the instructor only:

Problem abstract: Curves are one of the glories of elementary mathematics. Calculus students encounter lines and the conics but the next most complicated case, that of cubics, is rarely dealt with. One reason, of course, is that the cubic case is much more complicated, but a good deal of the difficulty can be swept away with the enlightened use of a computer. The goal of this project is to examine Newton's classification of cubics in so far as this is reasonable to do with elementary algebra, a computer, and some thought. The student will deal with a complex problem which has a pleasing solution involving many pretty pictures. For a complete catalog of the possible graphs (except for the six which Newton missed) see his paper mentioned in the bibliography.

Prerequisite skills and knowledge: For parts 1, 2, 3 and much of 5, it suffices to have a knowledge of calculus at the curve sketching level together with reasonable facility with a symbolic manipulation computer package (or a graphing calculator). Part 4 requires nothing but high school algebra, but the calculations will make more sense if the student has seen the reduction of the general quadratic in two variables to the standard conics using rotations and translations. It would be very reasonable to assign only parts 1-3 and 5 in a first year calculus class.

Essential/useful library resources: none

Essential/useful computational resources: This project requires the use of a graphing and symbolic algebra package.

Example of an acceptable approach:

1. These curves are sketched below. For the case of the folium (Figure 4) some care is needed in solving for y in terms of x. The usual algebraic solution of a cubic in terms of roots involving complex numbers is not suitable for drawing a real graph. A good equation solver will recognize this "irreducible" case and employ a trigonometric solution. Most students will not have encountered this use of trigonometry before and are likely to ask for an explanation. Consult an older book on the theory of equations, such as [3], or a good mathematical encyclopedia. An alternative which may make better sense to more advanced students is to use $y = tx$ to put the folium in parametric form. Most graphing tools will produce the proper graph from the parametric form $x = \dfrac{3t}{t^3 + 1}$, $y = \dfrac{3t^2}{t^3 + 1}$.

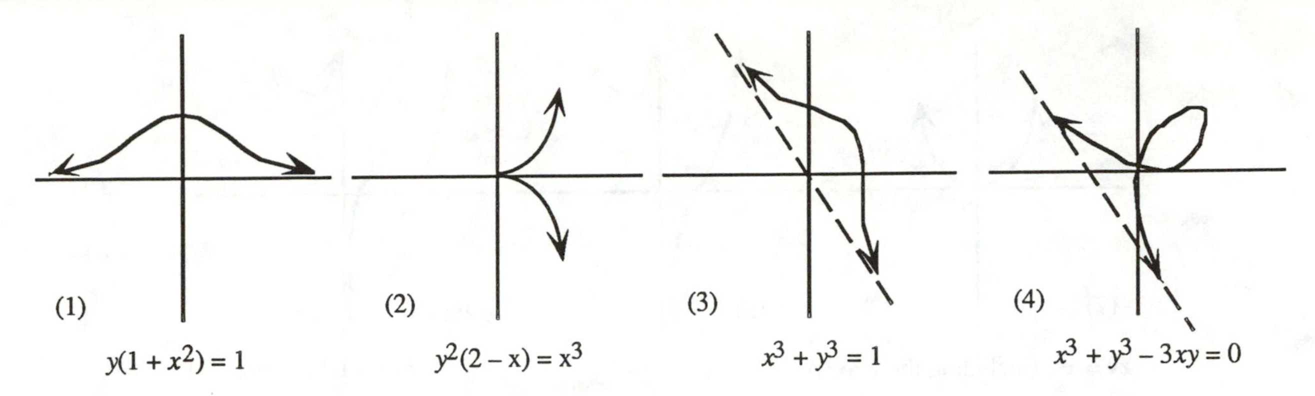

2. The shape is completely determined by the discriminant of $f'(x)$, $D = 4b^2 - 12ac$. The curve has two humps if $D > 0$ (Figure 6) and no humps otherwise (Figure 5). The sign of a controls the direction in which the curve approaches $+\infty$. Notice that we require that $a \neq 0$ so as to have a legitimate cubic. This curve is sometimes known as a cubical parabola.

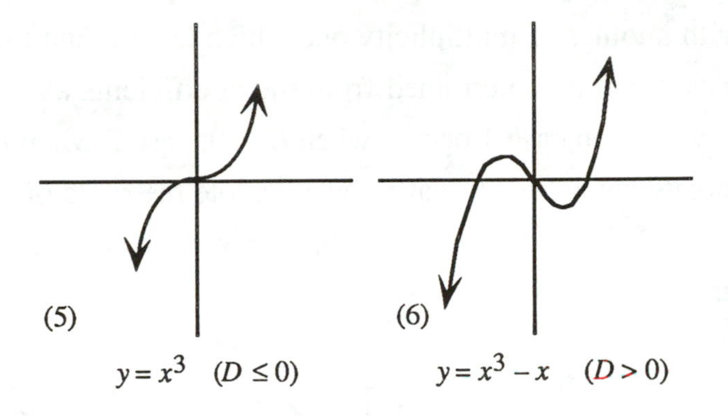

3. For this class of cubics there are three general shapes depending on d and the number of positive or negative roots of the equation $f(x) = 0$. If $d = 0$ the graph is the union of a parabola and the y-axis (Figure 7). If $d \neq 0$ the graph is known as the trident of Descartes or Newton. It has an asymptote at $x = 0$, two humps if $f(x) = 0$ has three positive or three negative roots (Figure 9), and no humps otherwise (Figure 8). It is interesting to find conditions on the coefficients which guarantee this root condition. First note that a necessary condition is that the discriminant D of f' is positive. Let $r < R$ denote the roots of $f'(x)$ (when $D > 0$). Then $f(x)$ will have three real roots if $f(r)$ and $f(R)$ have opposite signs. Another approach is to use the discriminant of the cubic f, which is discussed in [3]. In this case the roots will have the same sign if $r > 0$ and d and $f(r)$ have the opposite signs or if $R < 0$ and d and $f(R)$ have opposite signs. The sign of a, of course, has the same effect on the graph as in problem 2 above.

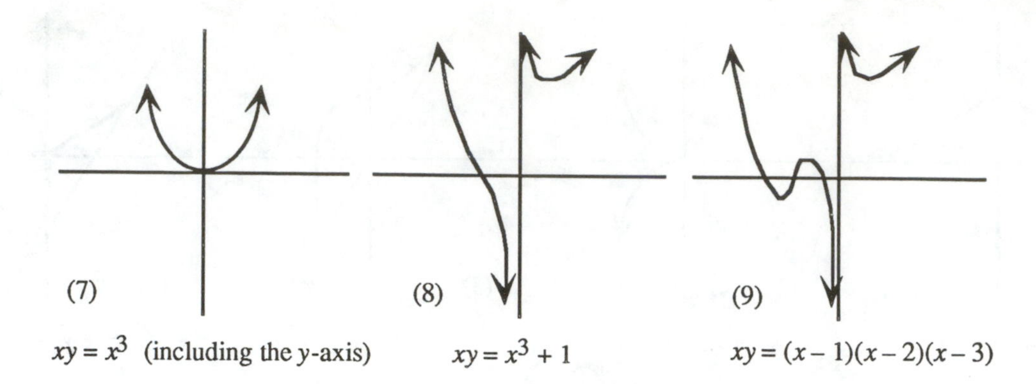

4. The shapes are classified by the number and location of the roots of $f(x)$. There are five cases: case 1: 1 real and 2 complex roots (Figure 10), case 2: 1 real root of multiplicity three (Figure 11), case 3: three distinct real roots (Figure 12), case 4: 1 real root of multiplicity two with another of multiplicity one which is greater than the root of multiplicity two (Figure 13), case 5: 1 real root of multiplicity two with another of multiplicity one which is less than the root of multiplicity two (Figure 14). These five cases can be determined from the coefficients as follows. We use the same notion as in problem 3 above. Then case 1 occurs when $D < 0$; case 2 when $D = 0$; case 3 when $D > 0$ and $f(r)$ and $f(R)$ have opposite signs (recall that r and R denote the roots of $f'(x)$); case 4 when $D > 0$ and $f(r) = 0$; and case 5 when $D > 0$ and $f(R) = 0$. Simple examples of the five situations are indicated below the appropriate figure.

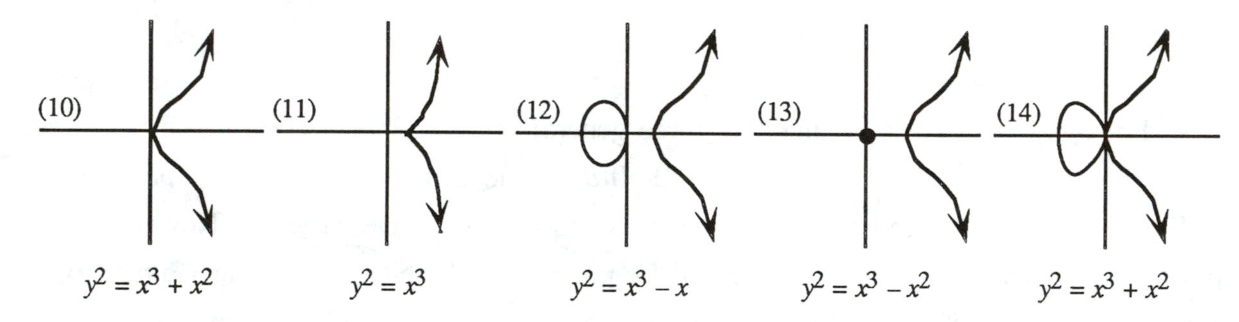

Figure 11 is known as the semicubical parabola. The graphs in Figures 10 and 13 look quite similar, but Figure 13 has an isolated point in addition to the main part of the curve.

5. This needs no explanation.

6. There are many species of cubic curves of type d) (71 in all). Newton studied these curves by completing the square to obtain the equation

$$(xy + e/2)^2 = g(x) = ax^4 + bx^3 + cx^2 + \mathrm{d}x + e^2/4. \qquad (2)$$

He then divided these curves into classes according to the sign of the leading coefficient a and further broke the classes into genera and species according to the behavior of the roots of $g(x) = 0$. We give some representative examples below. We obtained some of these examples by choosing simple values for the coefficients in equation (2) and others by selecting particular values for parameters in Newton's paper [4]. In some cases drawing a good graph requires zooming out to get the big picture or zooming in to see details. It is easy to miss a branch of the curve or an asymptote if the wrong scale is used. See Figure 23 below.

None of these curves can have more than two points above any given x value, since the original curve is quadratic in y. There are intervals on the x-axis with no corresponding (real) y values when $g(x) < 0$. This can happen on at most three intervals, since g has at most four real roots.

The degenerate cases (which are not counted in the 78 species) are any union of a conic with a line, a point with a line, or of three lines, some of which may coincide. Two of these are drawn below as Figures 25 and 26. Notice that a single point, which is a degenerate case for a quadratic curve, is not one for a (legitimate) cubic. Neither is the null set.

Any cubic of type d) will have from one to three asymptotes. Finding the asymptotes of a general algebraic curve seems to be a lost art. One needs to look in an older book, such as [5], to find the theorem that completely specifies them. Suppose that the curve is defined by an nth order polynomial equation $f(x, y) = 0$, and let $f_k(x, y)$ denote the pure k-th order part of f so that $f = f_n + f_{n-1} + \ldots$. If f has $ax + by$ as a simple factor so that $f_n = (ax + by)g_{n-1}$, then the curve $f = 0$ has the asymptote

$$(ax + by)g_{n-1}(b, -a) + f_{n-1}(b, -a) = 0.$$

If f_n has $ax + by$ as a repeated factor and $ax + by$ is also a factor of f_{n-1} so that $f_n = (ax + by)^2 g_{n-1}$ and $f_{n-1} = (ax + by)h_{n-2}$, then the curve has the pair of asymptotes

$$(ax + by)^2 g_{n-1}(b, -a) + (ax + by)h_{n-1}(b, -a) + f_{n-2}(b, -a) = 0.$$

Finally, if f_n has $ax + by$ as a repeated factor but f_{n-1} does not have $ax + by$ as a factor, then there is no asymptote of the form $ax + by = c$ for any constant c. Notice that the top three f_k's are necessary for a complete determination of the asymptotes, but the very top one f_n suffices to specify the (at most n) possibilities. The proof of this theorem relies on the notion of points at infinity—a line is an asymptote to a curve if and only if it is tangent to the curve at infinity. See [5, p. 7]. In the cubic case, at most three asymptotes are possible.

Newton's full classifications scheme shows that it is not possible to have more than four branches in any cubic curve. Examples of curves with these four possibilities are given below along with examples of other interesting shapes which occur for cubics of type d). Compare Figures 21 and 22 and Figures 24 and 25. In Figure 22 it is necessary to use several digits of accuracy for $\sqrt{8}$ to obtain the graph shown, which has a node.

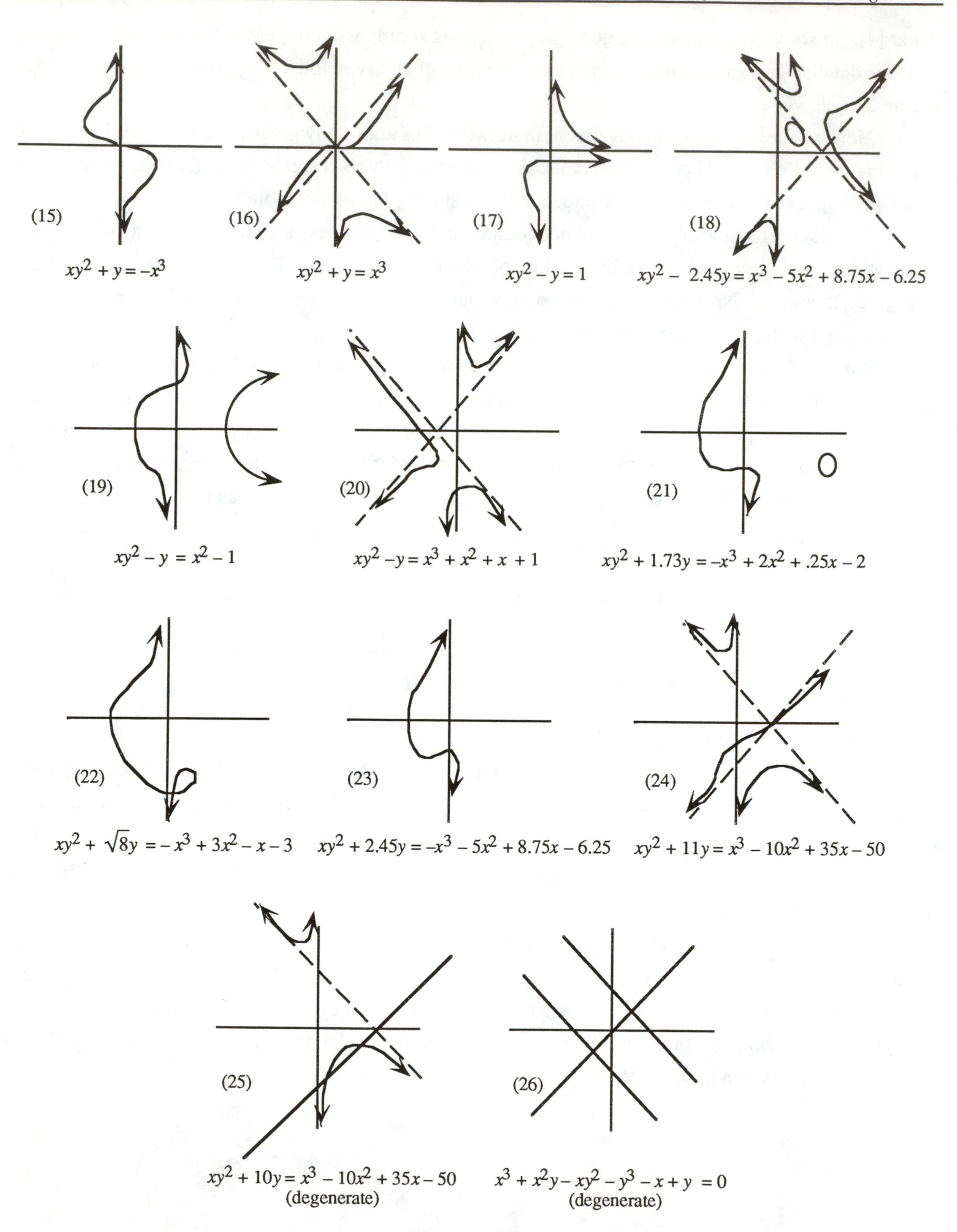
(15)
$xy^2 + y = -x^3$
(16)
$xy^2 + y = x^3$
(17)
$xy^2 - y = 1$
(18)
$xy^2 - 2.45y = x^3 - 5x^2 + 8.75x - 6.25$
(19)
$xy^2 - y = x^2 - 1$
(20)
$xy^2 - y = x^3 + x^2 + x + 1$
(21)
$xy^2 + 1.73y = -x^3 + 2x^2 + .25x - 2$
(22)
$xy^2 + \sqrt{8}y = -x^3 + 3x^2 - x - 3$
(23)
$xy^2 + 2.45y = -x^3 - 5x^2 + 8.75x - 6.25$
(24)
$xy^2 + 11y = x^3 - 10x^2 + 35x - 50$
(25)
$xy^2 + 10y = x^3 - 10x^2 + 35x - 50$
(degenerate)
(26)
$x^3 + x^2y - xy^2 - y^3 - x + y = 0$
(degenerate)

Questions for further exploration:

I. The following could be added to problem 5:

5.b) Obtain Newton's reduction to the four simpler types as follows. Rename x, y as u, v to avoid confusion with the coming transformations and consider the function defining (1)

$$h(u, v) = au^3 + bu^2v + cuv^2 + dv^3 + eu^2 + iuv + jv^2 + mu + nv + p,$$

where at least one of a to d is nonzero. Let $k(u, v) = au^3 + bu^2v + cuv^2 + dv^2$ denote the purely cubic portion of h. It is reasonable to work first with k, since it determines what the possible asymptotes are (although lower degree terms affect whether the possibilities are indeed asymptotes), and Newton based his classification on this asymptotic behavior.

Show that $k(u, v)$ is always divisible by some linear term $mu + nv$, where m and n are real constants, such that either

(i) $k(u, v) = (mu + nv)^3$, or

(ii) $k(u, v) = (mu + nv)(ru^2 + suv + tv^2)$, for real constants, r, x, and t with $mu + nv$ not dividing the second factor.

Let $x = mu + nv$ be a new coordinate. In case (i) let $y = pu + qv$ be any other coordinate which is independent of x, i.e., so that it is possible to solve for u and v in terms of x and y. Then in the x, y coordinate system $k(u, v)$ reduces to x^3. In case (ii) again let $y = pu + qv$ but show that a particular choice of p and q results in a coordinate system x, y so that $k(u, v)$ reduces to $xy^2 - ax^3$ for some real constant a.

The final step in the reduction is to make some of the other terms in h vanish. Show that further linear transformations or translations can be performed to arrive at a), b), or c) in case (i) and d) in case (ii). This proves that the general cubic equation can be reduced to one of these four simplified types.

Solution:

The algebra here can be extremely unpleasant if approached blindly. But if tackled from the proper point of view, this simplification can be performed with only a few elementary calculations. We follow the approach of Brieskorn in [2, p. 91]. Students may well need some additional explanation for this part beyond that written in the problem statement. To achieve this considerable simplification of the general cubic, Newton allowed a general affine change of coordinates. The orthogonal transformations which can be used to reduce conics to their standard forms do not suffice to simplify cubics.

We begin by considering the pure cubic form $k(u, v)$. To see that $k(u, v)$ has the stated factorization, set $z = u/v$. (If v is absent, this first part of the reduction is unnecessary.) There is at least one real root z of the resulting cubic in one variable, so $u - zv$ divides $k(u, v)$ by the factor theorem. If it divides it once or three times, this factor will do as $mu + nv$. If it divides $k(u, v)$ twice, use the linear quotient as $mu + nv$.

When k is a perfect cube (case i)), y can be chosen to be any other linearly independent coordinate, but in case ii) a calculation is necessary to determine an appropriate y to yield the result $xy^2 - ax^2$. We set y equal to $pu + qv$ and compute $y^2 - ax^2$ in terms of u and v. Here p, q, and a are to be determined. For $y^2 - ax^2$ to be equal to $ru^2 + suv + tv^2$, three equations need to be satisfied: $p^2 - am^2 = r$, $pq - mna = s/2$, and $q^2 - an^2 = t$. Then $p = \pm\sqrt{r + am^2}$ and $q = \pm\sqrt{t + an^2}$ (temporarily we allow p and q to be complex), so that $\pm\sqrt{r + am^2}\sqrt{t + an^2} - mna = s/2$. Squaring gives $(r + am^2)(t + an^2) = (s/2 + mna)^2$, hence $(m^2t + n^2r)a - mnsa = s^2/4 - rt$. This equation can be solved for a so long as $m^2t + n^2r - mns \neq 0$. But this condition is equivalent to the requirement that $mu + nv$ does not divide $ru^2 + suv + tv^2$. To see that p and q can be chosen to be real, notice that $pq - mna = s/2$ implies that one of p or q is real if the other is. Otherwise, both are pure imaginary, but in that case a factor of -1 can be removed from $ru^2 + suv + tv^2$ and included in the $mu + nv$ term, which now makes p and q real. Thus we have shown that appropriate real constants p, q, and a always exist in case ii). Hence there is always a suitable linear change of coordinates to simplify the pure cubic part of equation to one of the two forms ax^3 (in our approach a turns out to be 1) or $xy^2 - ax^3$.

To arrive at Newton's four types, another transformation must be performed to remove some of the lower degree terms. If $f(x, y) = ax^3 + py^2 + qxy + ry$ plus lower degree terms in x, then let $x' = x - m$ and $y' = kx + y - n$. Direct substitution into f shows that if $p \neq 0$ then suitable choices of m, k, and n can be made to cause the $x'y'$ and y' terms to vanish, which produces an equation of type c). If $p = 0$ but $q \neq 0$, then y' can be made to disappear, which yields type b). If both p and q are zero, the equation is already of type a). On the other hand, if $f(x, y) = xy^2 - ax^3 + py^2 + qxy + ry$ plus terms of lower degree in x, then a translation $x' = x - m$, $y' = y - n$ can be chosen to remove the y' and $x'y'$ terms. Thus we conclude that it is always possible by an affine transformations to reduce the general cubic (1) to one of the four types a) to d).

II. Special plane curves have always had an enthusiastic following among amateur and professional mathematicians. An attractive treatment of several special curves can be found in Simmons [7]. In particular, note his treatment of the folium of Descartes as a parametrized curve and as a polar curve (pp. 512, 548). These techniques, together with simply solving for y as a function of x, give three useful and complementary approaches to the study of this beautiful curve. Note also Simmons' description of how certain special curves can be used to solve famous geometric problems of antiquity (p. 493). Other good references for special curves are Stillwell [8] and Brieskorn [2]. A useful order treatment of curves, which also partly classifies quartic curves, is Salmon [6].

It is possible for an industrious student to examine Newton's complete classification of cubics. The best way to do this is to look at Ball's commentary [1], which explains what is going on in (somewhat) modern language and at the same time to look at the pictures in Newton's paper [4].

Newton's criteria for the classification are the nature of the asymptotes and the extremal behavior of the curves. This scheme was criticized by later authors, since it resulted in so many different cases. Actually, Newton showed a way to simplify the classification with an (unproved) remark in his paper.

He claimed and later authors proved that each cubic is a projective image of one of the five species of cubic of type c). In more modern language his assertion is that every irreducible cubic is isomorphic as a curve in real projective space to one of these five species. Working in projective space amounts to adding points at infinity and enlarging the group of permissible transformations from the affine group, which Newton used, to an appropriate larger one, the projective linear group. It has been noted earlier that adding points at infinity allows one to deal with asymptotes as ordinary tangent lines. It is also clear that branches of a curve which share a common asymptote will be joined with the addition of these extra points, so this device will reduce the number of branches.

Another simplification occurs if the curves are viewed as complex curves, or putting both ideas together, as curves in complex projective space. In this more general space the number of distinct types of irreducible cubics reduces to three, one with a cusp, one with a node, and the third nonsingular. Newton's study dealt with the finite real portions of these curves. By looking at the "entire curve" later mathematicians achieved this tremendous simplification.

All three types of complex projective cubic curves turn out to be connected; i.e., they have exactly one branch. In fact, topologically they are either a sphere, a sphere with two points identified, or a torus. In real projective space a cubic curve has either one or two branches, as can be seen by examining the five species of type c). In the ordinary real plane Newton's full classification scheme shows that no cubic curve can have more than four branches. I know of no way to deduce this maximum of four without examining the 78 cases.

A good general reference for the ideas discussed in the last several paragraphs is Stillwell [8] (see pages 69, 80, 208, 231). For an excellent modern but reasonably elementary discussion of plane curves, projective geometry, and the beginnings of algebraic geometry, see Brieskorn [2].

References/bibliography/related topics:

1. Ball, W. W. R.: *Newton's classification of cubic curves*, *Proceedings of the London Math. Society* 22, 104-143 (1890).
2. Brieskorn, Edgert and Knörrer, Horst: *Plane Algebraic Curves*, Birkhäuser, Boston, 1986.
3. Dickson, L. E.: *New First Course in the Theory of Equations*, Wiley, New York, 1939.
4. Newton, Isaac: *The Mathematical Works of Isaac Newton*, edited by Derek Whiteside, vol. 2 and 7, Cambridge University Press, 1967-1981.
5. Primrose, E. J. F.: *Algebraic Plane Curves*, Macmillan, London, 1955.
6. Salmon, George: *A Treatise on the Higher Plane Curves*, Hodges, Forster, and Co., Dublin, 1873.
7. Simmons, George: *Calculus with Analytic Geometry*, McGraw Hill, New York, 1985.
8. Stillwell, John: *Mathematics and Its History*, Springer Verlag, New York, 1989.

Title: Cavalieri's Integration Method

Author: Mic Jackson, Earlham College

Problem Statement: Bonaventura Cavalieri (1598-1647), a student of Galileo, was one of the key figures in the development of infinitesimal methods. His techniques were based upon the work of Archimedes and provided ideas and methods which were important in the later development of the infinitesimal calculus by Newton and Leibniz. You will need to read this document carefully, being sure that you understand why each statement is true, given Cavalieri's assumptions. I have asked you to demonstrate why certain statements are true. To **demonstrate** in this context means to use geometry, algebra, and/or deductive logic to provide a conclusive argument that a certain statement is true. You will need to read carefully, think deeply and test your ideas. It will help your understanding of Cavalieri's methods if you try to relate them to your own understanding of the use of Riemann sums in determining areas. When you are finished you should have a greater appreciation of the power and sophistication of the modern definition of the definite integral.

In addition, in your final report, you should describe any questions which you have not been able to resolve and any new insights you have gained through studying Cavalieri's work. Your report should not be more than 6 double-spaced pages, including diagrams and bibliography. Your work will be evaluated for mathematical correctness, thoroughness and clarity, as well as for the standard criteria of any written submission. You may need to use library resources; be sure to cite those you do use.

Cavalieri's Integration Method:

(Adapted from *The Origins of the Infinitesimal Calculus*, by M. E. Baron, pages 122-135.[1])

In most of his work, Cavalieri *assumed* that a planar region was made up of an indefinite number of equidistant parallel straight lines and that a solid was made up of an indefinite number of equidistant parallel planes. He sometimes used the analogies of a garment being made up of parallel threads woven together or of a book being comprised of parallel leaves. For plane (solid) figures he defined a *regula* to be a line (plane) drawn through a convenient vertex parallel to a "base" as the starting point of his integration. In Figure 1 below, the base of the planar region is the line *BC* and the *regula* is the line *EO*. The related planes form the base and *regula* in Figure 2. He then imagined moving the *regula* parallel to itself until it came into coincidence with the base of the object. He treated the intercepts (lines or plane sections) of the *regula* with the original plane (or solid) figure as indivisibles which "taken together" form the planar region or solid.

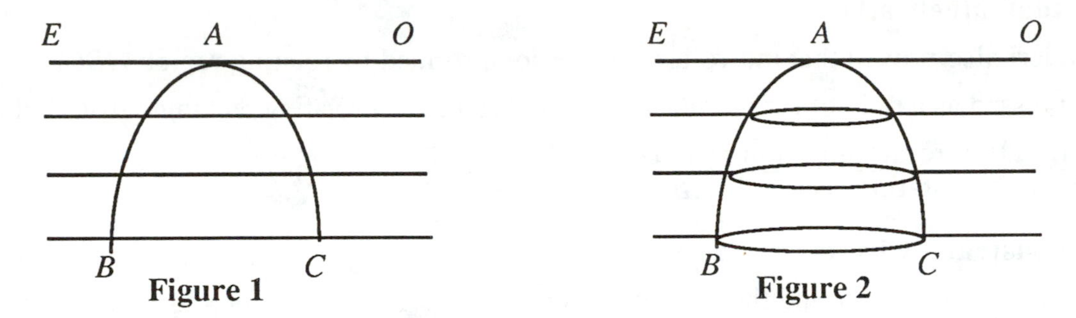

Figure 1 **Figure 2**

In his systematic integration method, Cavalieri derived the relation between the areas (or volumes) of two planar regions (or solids) from the ratio of the summations of the line (or surface) indivisibles. In doing so, he made use of the notion of the infinite only in an subsidiary way. His technique did not apply to all planar or solid shapes, but only to those shapes that could be described by what we would call a power function. He used the concept of *powers of lengths of line elements* to reduce the process of calculating area or volume to that of finding the sums of the powers of the lengths of line elements in a triangle taken parallel to a given *regula*. Consider Figure 3.

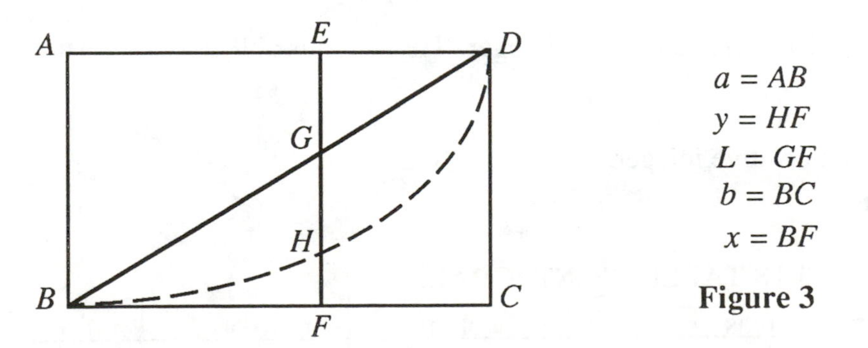

Figure 3

If we take segment *AB* as the *regula* and segment *CD* as the base of rectangle *ABCD*, then let *EF*, drawn parallel to *AB*, cut the curve in *H* and the diagonal *BD* in *G*. Then according to Cavalieri's assumption:

---

[1] Adapted with the permission of Macmillan Publishing Company, a Division of Macmillan, Inc. from THE ORIGINS OF THE INFINITESIMAL CALCULUS by Margaret Baron. Copyright ©1969 Pergamon Press.

$$\frac{\text{area of region } HBCD}{\text{area of region } ABCD} = \frac{\sum HF}{\sum AB} = \frac{\sum y}{\sum a}.$$

(1. Explain how this follows from Cavalieri's assumption.)

To understand what is meant by the powers of lengths of line elements, assume the curve *BHD* in Figure 3 is of the form $\frac{y}{a} = \left[\frac{x}{b}\right]^n$ with the origin at point *B* and other magnitudes named as shown.

Recognition of certain similar triangles then allows us to say that $\frac{y}{a} = \left[\frac{L}{a}\right]^n$

(2. Show that the last statement follows from the assumption of the form of curve *BHD*.)

After some careful algebra we can conclude that the $\frac{\text{area of region } HBCD}{\text{area of region } ABCD} = \frac{\sum L^n}{\sum a^n}$.

(3. Do that algebra.)

Cavalieri also proved that the volume of the solid formed by rotating region *HBCD* about line *BC* can be expressed as a ratio of the volume of the solid formed by rotating rectangle *ABCD* about line *BC*:

$$\frac{\text{solid formed by rotating region } HBCD}{\text{solid formed by rotating rectangle } ABCD} = \frac{\sum y^2}{\sum a^2} = \frac{\sum L^{2n}}{\sum a^{2n}}.$$

(4. Demonstrate why this is so.)

Since he knew how to find the area of rectangle *ABCD* (and the volume of the cylinder formed by rotating rectangle *ABCD* about line *BC*), Cavalieri could find the area of this planar region *HBCD* defined by any of these special curves (and the volume of the solid formed by rotating region *HBCD* about line *BC*) by using the sums of the powers of the lengths of the line elements of triangle *BCD*. Using a rather ingenious approach that involved geometric notions of congruence, similarity and equality, and algebraic results with binomial coefficients, he was able to determine the sums of the powers of lines of a triangle up to and including $\sum L^9$ (See Table 1 below). He concluded that $\frac{\sum L^n}{\sum a^n} = \frac{1}{n+1}$ if $a = 1$ and n is a positive integer. He was thus able to draw up a table of integrals of the form $\int_0^1 x^n dx$ for n a positive integer.

PART OF CAVALIERI'S TABLE OF INTEGRALS:

function	ratio of areas	ratio of volumes	ratio of ?	ratio of ??
$y = x$	$\frac{1}{2}$	$\frac{1}{3}$	$\frac{1}{4}$	$\frac{1}{5}$
$y = x^2$	$\frac{1}{3}$	$\frac{1}{5}$	$\frac{1}{7}$	$\frac{1}{9}$
$y = x^3$	$\frac{1}{4}$	$\frac{1}{7}$	$\frac{1}{10}$	...

Interpreted in ways we can understand, the first column of Cavalieri's table of integrals gives the proportion of the area of the unit square taken up by the region to the right of and below the curve of $y = x^n$. (This would be curve *BHD* in Figure 3.) Since we know the area of the unit square we can use the table to find the area of a class of planar regions. The second column gives the proportion of the volume of the solid region form by rotating the lower right region about the line $y = 0$ to the volume of the cylinder formed by rotating the unit square about the line $y = 0$. Since we know how to calculate the volume of a cylinder we can find the volume of a class of solids of revolution. Columns beyond the first two have no physical interpretation.

CAVALIERI'S INGENIOUS APPROACH

We will reconstruct Cavalieri's proof that $\frac{\sum L^5}{\sum a^5} = \frac{1}{6}$, assuming that he has already shown that $\frac{\sum L^n}{\sum a^n} = \frac{1}{n+1}$ for $n = 1, 2, 3$ and 4. Consider the parallelogram *ABCD* (Figure 4); divide it into two congruent triangles with diagonal *BD* and into four congruent parallelograms by lines *EF* and *GH* which meet at *M*. Then draw segments *PQRS* and *P'R'Q'S'* parallel to and equidistant from *EF*. Then let $PQ = Q'S' = a$, $QR = Q'R' = b$ and $RS = R'P' = a - b$.

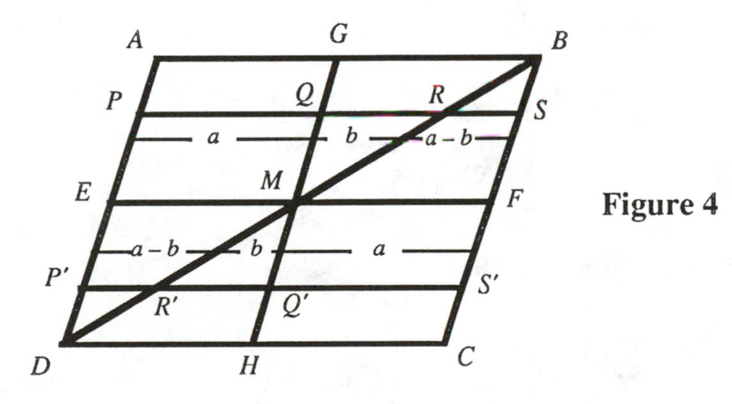

Figure 4

$\sum_{ABME} (a+b)^5 + \sum_{BFM} (a-b)^5$ is the summation of the fifth-powers of the line indivisibles making up the top half of *ABCD*. It is true that $(a+b)^5 + (a-b)^5 = 2a^5 + 20a^3b^2 + 10ab^4$.

(5. Demonstrate that this last equation is true.),

It follows that $\sum_{ABME} (a+b)^5 + \sum_{BFM} (a-b)^5 = \sum_{AGME} 2a^5 + \sum_{GBM} 20a^3b^2 + \sum_{GBM} 10ab^4.$

(6. Explain why this equation makes sense.)

Similarly for the bottom half of *ABCD*,

$$\sum_{MFCD}(a+b)^5 + \sum_{EMD}(a-b)^5 = \sum_{MFCH}2a^5 + \sum_{DMH}20a^3b^2 + \sum_{DMH}10ab^4.$$

Considering the entire parallelogram $ABCD$, given that parallelogram $AGME$ is congruent to parallelogram $MFCH$ and that triangles GMB and HMD are also congruent, it follows that

$$\sum_{ABCD}2PR^5 = \sum_{AGHD}2a^5 + \sum_{GBM}40a^3b^2 + \sum_{GBM}20ab^4.$$

(7. Demonstrate that this last equation is true.)

Because he had already shown that $\dfrac{\sum L^n}{\sum a^n} = \dfrac{1}{n+1}$ for $a = 1$ and $n = 1$ through 4, Cavalieri was able to conclude that $\sum_{GBM} b^2 = \frac{1}{3}\sum_{GBFM} a^2 = \frac{1}{6}\sum_{GBCH} a^2$ and $\sum_{GBM} b^4 = \frac{1}{10}\sum_{GBCH} a^4.$

(8. Explain why these equations make sense.)

From this it follows that $\sum_{ABCD} 2RS^5 = \sum_{AGHD} 2a^5 + \frac{40}{6}\sum_{GBCH} a^5 + \frac{20}{10}\sum_{GBCH} a^5 = \frac{32}{3}\sum_{AGHD} a^5.$

(9. Demonstrate that this last equation is true.)

The sum of the fifth powers of the line indivisibles for the entire parallelogram is $\sum_{ABCD} AB^5$. Since $AB = 2a$, then $AB^5 = 32a^5$ and $\sum_{ABCD} AB^5 = 32\sum_{AGHD} a^5$, it quickly follows that $\sum_{ABCD} RS^5 = \frac{1}{6}\sum_{ABCD} AB^5.$

(10. Demonstrate that this is true.)

Cavalieri thus proved that $\dfrac{\sum L^5}{\sum a^5} = \dfrac{1}{6}$ for a and L defined as shown in Figure 5.

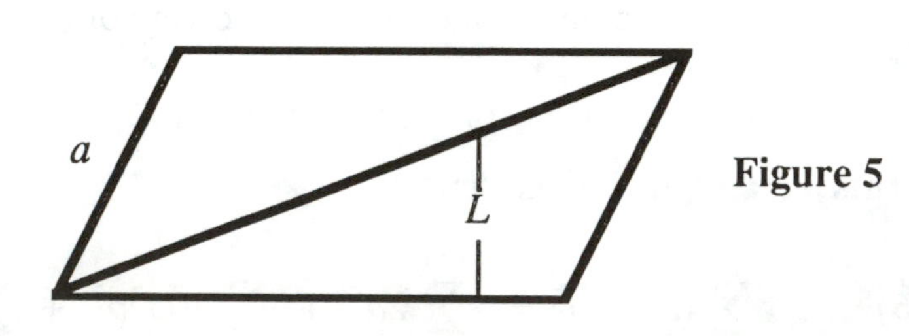

Figure 5

Information for the instructor only:

Problem abstract: This library research and writing project provides small groups of students the opportunity to wrestle with one of the difficult problems of science; that of whether matter and time are made up of "smallest" pieces or infinitely divisible. Cavalieri's indivisible techniques are based upon two distinct and complementary approaches which he designated by the terms *collective* and *distributive*. Under the collective approach the sums, $\sum L$ and $\sum M$, of the line (or surface) indivisibles for two figures $P1$ and $P2$ are first obtained separately and then used to establish the ratio of the areas (or volumes) of the figures themselves. If, for example, $\frac{\sum L}{\sum M} = \frac{a}{b}$, then $\frac{\text{measure of } P1}{\text{measure of } P2} = \frac{a}{b}$. This is the method employed in this project.

This project does not address the well-known theorem ascribed to Cavalieri, but the students may run into it in other sources, so it is important that they realize that the principles stated in this theorem are not employed in the method discussed here. Cavalieri's principles are fundamental to the "distributive technique," which "was developed primarily in order to meet the philosophic objections which Cavalieri felt might be raised against the comparison of indefinite numbers of lines and planes." (Baron, page 126) These principles are: (1) areas enclosed by two planar regions are equal provided that any system of parallel lines cuts off equal segments in each; and, (2) volumes enclosed by two solid figures are equal provided that any system of parallel planes cuts off equal planar regions in each. A corollary of these principles is that if for every pair of corresponding intercepts L and M, $\frac{L}{M} = \frac{a}{b}$, then $\frac{\text{measure of } P1}{\text{measure of } P2} = \frac{a}{b}$. Cavalieri only used this method in a small number of cases where $\frac{a}{b}$ is constant for each pair of intercepts. However, mathematicians in the seventeenth century developed this method into a valuable means of integration by geometric transformation.

Prerequisite skills and knowledge: The students will need to know geometric formulas for area and volume of some basic figures, high school algebra, the binomial theorem, and should have had an introduction to the Riemann definition of the definite integral.

Essential/useful library resources: See the bibliography in the introduction to Historical Projects.

Essential/useful computational resources: none

Example of an acceptable approach:

1. Explain how $\frac{\text{area of region } HBCD}{\text{area of region } ABCD} = \frac{\sum HF}{\sum AB} = \frac{\sum y}{\sum a}$ follows from his assumption.

He considered each planar region to be made up of its line indivisibles. Hence, each area is simply the sum (in some infinite sense) of the line indivisibles. The equal ratios follow from the fact that if $x = y$ and $m = n$ then $\frac{x}{m} = \frac{y}{n}$.

2. Show that $\frac{y}{a} = \left[\frac{L}{a}\right]^n$ follows from the assumption of the form of the curve *BHD*.

Triangle *BGF* is similar to triangle *BCD*; hence, $\frac{x}{b} = \frac{L}{a}$. So $\frac{y}{a} = \left[\frac{x}{b}\right]^n = \left[\frac{L}{a}\right]^n$.

3. Do the algebra to show that $\frac{\text{area of region } HBCD}{\text{area of region } ABCD} = \frac{\sum L^n}{\sum a^n}$.

Remembering that "a" is constant, the given relationship allows us to see that, for the kth line indivisible, $\frac{y_k}{a} = \left[\frac{L_k}{a}\right]^n = \frac{(L_k)^n}{a^n}$. Now the question is whether a summation of all of those line indivisibles will preserve the equality. Note the likely student error of arguing that $\sum_{k=1}^{\infty} \frac{y_k}{a} = \sum_{k=1}^{\infty} \frac{(L_k)^n}{a^n}$, which is true, but not what we need. (I am being a bit cavalier with my infinite series here, but so was Cavalieri!) We have to approach it a bit differently. Since $\frac{y_k}{a} = \frac{(L_k)^n}{a^n}$ it follows that $y_k = a\frac{(L_k)^n}{a^n} = \frac{(L_k)^n}{a^{n-1}}$ holds for each line indivisible.

Hence, $\frac{\sum y}{\sum a} = \frac{\sum \frac{L^n}{a^{n-1}}}{\sum a} = \frac{\sum L^n}{a^{n-1}\sum a} = \frac{\sum L^n}{\sum a^n}$.

4. Demonstrate why $\frac{\text{solid formed by rotating region } HBCD}{\text{solid formed by rotating rectangle } ABCD} = \frac{\sum y^2}{\sum a^2} = \frac{\sum L^{2n}}{\sum a^{2n}}$ is true.

According to the argument in 3, the ratio of areas of the circles for the kth indivisibles, which will be the surface indivisibles in this case, will be $\frac{\pi y_k^2}{\pi a^2} = \frac{y_k^2}{a^2} = \frac{(L_k)^{2n}}{a^{2n}}$. The rest also follows from 3.

5. Demonstrate that $(a+b)^5 + (a-b)^5 = 2a^5 + 20a^3b^2 + 10ab^4$ is true.

If the students will simply expand the binomials and combine like terms, this works out pretty easily.

6. Explain why $\sum_{ABME} (a+b)^5 + \sum_{BFM} (a-b)^5 = \sum_{AGME} 2a^5 + \sum_{GBM} 20a^3b^2 + \sum_{GBM} 10ab^4$ makes sense.

This seems a bit strange, but I guess that's why mathematicians have considered it ingenious. I would hope that students would argue along the following lines. The relationship justified in argument 5 is true for each line indivisible in parallelogram *ABFE*. So if we add *all* the line indivisibles, the equality of those sums must also hold.

Now why did Cavalieri make the summations over only the regions shown? Clearly $ABME$ is made up of line indivisibles of the form $(a + b)$ and BFM is made up of line indivisibles of the form $(a - b)$. On the right side of the equality, there are no magnitudes $(a + b)^5$ or $(a - b)^5$, so we don't need to sum over regions $ABME$ or BFM. In the planar region $AGME$, only line indivisibles of length "a" are present, so it is reasonable to sum the a^5 term only there. In triangle GBM, only line indivisibles of length "b" are present, but "a" is a constant magnitude, so it is reasonable to sum products of powers of "a" and "b" in GBM.

7. Demonstrate that $\sum_{ABCD} 2PR^5 = \sum_{AGHD} 2a^5 + \sum_{GBM} 40a^3b^2 + \sum_{GBM} 20ab^4$ is true.

Adding the summations for regions $ABFE$ and $EFCD$, we get on the left side of our equation: $\sum_{ABME} (a + b)^5 + \sum_{BFM} (a - b)^5 + \sum_{CDMF} (a + b)^5 + \sum_{DEM} (a - b)^5$, which can be written as $\sum_{ABCD} (PR^5 + RS^5)$ because $ABME \cong CDMF$ and $BFM \cong DEM$. Further, since $ABD \cong CDB$, it follows that $\sum_{ABD} PR^5 = \sum_{CDB} RS^5$, and we can write the left side as $\sum_{ABD} 2PR^5$, which is the same as $\sum_{ABCD} 2PR^5$. If we combine the expressions for the

areas of the top and bottom of $ABCD$, on the right side of the equation we get

$$\sum_{AGME} 2a^5 + \sum_{GBM} 20a^3b^2 + \sum_{GBM} 10ab^4 + \sum_{MFCH} 2a^5 + \sum_{DMH} 20a^3b^2 + \sum_{DMH} 10ab^4.$$

Since $AGME \cong MFCH$ and $GBM \cong DMH$, it follows that we can write the right side as

$$\sum_{AGHD} 2a^5 + \sum_{GBM} 40a^3b^2 + \sum_{GBM} 20ab^4.$$

8. Explain why these make sense: $\sum_{GBM} b^2 = \frac{1}{3} \sum_{GBFM} a^2 = \frac{1}{6} \sum_{GBCH} a^2$ and $\sum_{GBM} b^4 = \frac{1}{10} \sum_{GBCH} a^4$.

Apparently to Cavalieri terms like "b^2" meant that one was to take each line indivisible, find the length "b" on it, and then square that magnitude. Then add up all those squares. I think his reasoning must have been something like this: For the first equation, the magnitude b^2 *averages* $\frac{1}{3}a^2$ over GBM (based on his earlier work that yielded the table in the narrative). Also, b^2 *averages* $\frac{1}{3}a^2$ over $GBFM$ since no b-length is added to any line indivisible by attaching triangle BFM. Finally, the area of $GBCH$ is twice that of $GBFM$. Similar reasoning allows the conclusion that b^4 *averages* $\frac{1}{5}a^4$ over $GBFM$ and $\frac{1}{10}a^4$ over $GBCH$.

9. Demonstrate that $\sum_{ABCD} 2RS^5 = \sum_{AGHD} 2a^5 + \frac{40}{6} \sum_{GBCH} a^5 + \frac{20}{10} \sum_{GBCH} a^5 = \frac{32}{3} \sum_{AGHD} a^5$ is true.

From argument 7, concerning the sum of a^3b^2 over GBM, a is constant, so a^3 is too. Hence, from argument 8, over *GBCH* a^3b^2 averages a^3 times $\frac{1}{6}a^2$ or $\frac{1}{6}a^5$. Likewise, over GBCH ab^4 averages $\frac{1}{10}a^5$. The second expression in the equation follows immediately. Finally, since $AGHD \cong GBCH$, we can conclude that

$$\sum_{ABCD} 2RS^5 = \frac{32}{3} \sum_{AGHD} a^5.$$

10. Demonstrate that $\sum_{ABCD} RS^5 = \frac{1}{6} \sum_{ABCD} AB^5$ is true.

This follows immediately from $\sum_{ABCD} 2RS^5 = \frac{32}{3} \sum_{AGHD} a^5$ and $\sum_{ABCD} AB^5 = 32 \sum_{AGHD} a^5$.

Conjectures we expect that some students will make:

(1) Students may accept indivisibles as "truth." Cavalieri's approach to them will seem equivalent to the Riemannian definition of the definite integral.

(2) Some may be troubled by questions such as those stated below. Encourage them to follow up on their ideas.

Questions for further exploration:

(1) Prove that $\frac{\sum L^n}{\sum a^n} = \frac{1}{n+1}$ for $n = 6$ using Cavalieri's method.

(2) How did Cavalieri resolve the problem of "jagged edges" if his indivisibles had any thickness?

(3) How *thick* were the lines/planes used to cut the planar/solid regions? Did Cavalieri perceive that they had some infinitesimal thickness, or did he assume that they could be arbitrarily thin?

(4) How did Cavalieri actually carry out infinite summations of indivisibles?

References/bibliography/related topics:

(1) See library sources above and in the introduction to Historical Projects.

(2) There are a number of related historical projects in this volume.

Special implementation suggestions:

See the implementation suggestions in the introduction to Historical Projects.